Appalachian Tales

Deanna Edens

Appalachian Tales

Acknowledgments

Special thanks to David Robert Edens Jr., Cheryl Estrada, Nancy Holloway, Barbara L. Jones, Geneva Lacy, Sue Pilski, and Pam Tindell for providing editing advice.

Other Books by Deanna Edens

The Convenience of Crafting Maple Fudge
Welcome to Bluewater Bay
Christmas Comes to Bluewater Bay
Mystery in Bluewater Bay
Love Blooms in Bluewater Bay
The Adventures of the Bluewater Bay Sequinettes:
The Complete Bluewater Bay Series
Angels of the Appalachians
Molly's Memoir
Erma's Attic: Angels of the Appalachians Book Two
Rosa's Castle
Jinx at the Greenbrier
Will's Journey: Angels of the Appalachians 3
The Almost Cool Kids Club
Tids and Bits: Illustrated Reflections on Daily Living
The Abolitionist's Wife
Sweet Springs: A Novella
Pearl: You are Cleared to Land

For Nadia

"Because every picture tells a story."

"Photography takes an instant out of time, altering life by holding it still."

— *Dorothea Lange*

Preface

This summer, while I was visiting my old stomping grounds in Charleston, West Virginia, I took a drive through the neighborhood where I lived for a short time in the early eighties. Noticing a house that had a "For Sale" sign pinned in the front yard, I pulled over and stared at the modest ranch-style home for a long spell. Nostalgic memories of the times I spent with my dear friend, Nadia, flooded over me.

I have always enjoyed listening to people's tales and the stories, and the photographs and newspaper clippings Nadia shared with me years ago have remained in my thoughts for years. Although she has passed on over to the sweet by and by, I believe she would approve of my sharing some of her life events with you. She and I spent two years as neighbors and during this time she shared numerous stories about her family, the Benwood Mine Disaster, and sagas about bigotry, integration of schools and unsolved mysteries. Nadia even told me tales about her husband, who was a newspaper reporter. He covered events throughout the state of West Virginia for decades.

Oh, and the photographs she had stored in her precious memory album were images that stay in my mind's eye forever! She explained that *her* album was called a *memory album* because it wasn't simply a photo album. It was more than that. It was a collection of snapshots that captured time, and she valued the moments—each and every one of them. When she viewed the photographs, she slipped back to the night the stars winked despondently in the midnight blue sky above them, or the steamy summer day when she and Isaac said their vows. There were dozens of memories confined within her cherished album.

When I returned from Charleston, I rummaged through old notebooks and a box of newspaper articles Nadia had given to me years ago. I recalled the story she told me about the mysterious and tragic disappearance of the Sodder children in 1945, and I found a snapshot of the smoldering house on the front page of an old, faded newsprint. The photo reminded me of how life changes in a moment—sometimes for the better and sometimes for the worse.

Ultimately, I decided to record Nadia's stories. *Appalachian Tales* is a portrait of a life that was packed full of history, love, heartbreak, acts of kindness, bravery, joy and mysterious events. With this being noted, I sincerely hope you enjoy reading about Nadia and her notable accounts as much as I enjoyed scribing them, because when I am writing I rely on this passage written by Rudyard Kipling, "If history were told in the form of stories, it would never be forgotten."

Warm Regards,
Deanna Edens

Charleston, West Virginia

1982

"You should have seen him in color." Nadia placed the black and white photograph down on the table. "He was, without a doubt, the best-looking man I ever laid my eyes on. You can't tell it here but his eyes were sage-green and his shock of curly hair was as gold as a maple tree in October."

"I can see he was handsome even through the shades of gray," I replied sincerely. "What year was this photo taken?"

"1924."

"He had some big, broad shoulders, eh?"

Nadia nodded musingly. "Yes, he did. I'm a shoulder kind of gal. You know some women like firm rumps and some are attracted to muscular arms, but not me. I've always liked men with broad shoulders." Nadia drummed her fingers on the table. "You know Dee, a good man is as hard to find as the perfect fitting bra," she advised with utmost certainty. She tucked the aged snapshot back into the faded memory album.

"I wouldn't know, Nadia. I haven't found either one yet."

She chuckled and pointed to me. "Here's another bit of good advice. A lady should always wear clean undies just in case she might be in a wreck."

I peeped at her inquisitively. "Why's that?"

"You wouldn't want to go to the hospital with dirty underwear on. I mean what if they had to operate on ya? Huh? What would the doctors and the nurses think?"

"Really?" I arched my brow teasingly. "If I wreck my car I'm pretty sure my undies won't be clean by the time I arrive at the hospital."

She held her weathered hand up defensively. "I know. I know. You're twenty-some-years old now and know everything."

"I do not think I know everything, Nadia." I poured another glass of sweet tea that was as thick as syrup. "I just think that if I'm in an accident, I'll probably pee my britches."

"A lady doesn't talk about peeing in her britches either," she clucked her tongue. "I don't know what's wrong with the younger generation. When I was growing up we would never had used the word pee in public."

I opened my mouth to retort then prudently decided against it. "Sorry."

I knew that Nadia was born in 1906 in a three-room shack in West Virginia's coal country. Although she didn't have an abundance of money, she had what matters the most—respect for others and manners. She was a true Southern girl and knew to say please and thank ya, along with the other finer points of being a lady, such as never answering the door with curlers in your hair and, of course, putting your hands into your prayers. Nadia always said that a Southern lady 'hides her crazy with fine manners' and she had imparted her words of wisdom to her daughters, Sue and Sophia, and to any of her grandchildren who would take five minutes of their day to listen to her advice.

However, they didn't come around too often and since Nadia lived a few houses down from me I figured I would step in and be her surrogate grandchild. I liked Nadia. She was a petite, round lady with thick white curls, twinkling hazel eyes, and cheeks that shone like a pair of ripe cherries. She also had a gift for transforming something new and useful from discarded items. She could do anything with her hands. She painted tables, shelves and chairs that she found along the side of the road, turning them into beautiful works of art.

"Have I told ya the story about Mrs. Pruitt?" Nadia posed.

"I don't think so. Who's Mrs. Pruitt?"

"Mrs. Pruitt is Geraldine's sister," Nadia began as she walked to the sink, shoved the plug into the drain, and added a generous dollop of dishwashing liquid. I heard the grudging thunk of tired pipes before water gushed into the chipped porcelain sink.

"Who's Geraldine?"

"You don't know her."

Nadia always did this to me. She'd make a reference to somebody's sister's niece, or a brother of a father-in-law and I never knew who she was talking about. Most of the time she would continue on with her explanation of why and how they were kin to one another. Honestly, I would need a Venn diagram to follow along with her explanation and in the end I'd most likely be more confused, so I always dip my chin agreeably and murmur a little "Mmm, hmm."

"Anyway, since I'm not one to go 'round spreading rumors, ya better listen close the first time," Nadia informed me.

"I'm all ears."

"Well, Mrs. Pruitt was at church three Sundays ago and got up in the middle of services to go visit the ladies' room."

I walked over to the sink, bumped her with my bum, and insisted, "You go sit down. I'll wash up the dishes."

"Thank you, Dee." She ambled back over to the chrome-and-vinyl dinette table and dropped down into a chair. "Now, Mrs. Pruitt was not born with the grace of a gazelle, bless her heart, and when she returned from the bathroom and started walking up the aisle everyone could see that she had accidentally tucked the tail of her dress into her underwear." Nadia stifled a giggle. "Then, when Jack Kerns saw her exposing her backside he reached out from where he was seated in the pew and attempted to yank her dress out."

I pictured the scene in my mind, the carved pews worn smooth by at least a half century of parishioners sliding in and out of them, and the elderly Mr. Kerns attempting to help Mrs. Pruitt salvage some dignity. "I suppose that was the gentlemanly thing to do."

"Yep, probably so. But it startled Mrs. Pruitt and she turned and blinked at him like a possum whose hidey-hole had been unexpectedly exposed. She forcefully tugged at her attire causing her oversized, silky bloomers to drop to the floor."

"Did she mange to straighten up her dress before they hit the floor?"

"Yes, she did, and the entire congregation buried their heads in their hands and tried not to laugh."

"How embarrassing," I replied.

"It was for sure. She blushed four shades to Sunday before she scrunched down and attempted to discreetly wiggle her undies back up onto her hips."

"Were they clean?" I teased.

"They were spotless." Nadia nodded fervently.

Spotless? Seriously?

She continued, "Which is exactly my point. It could've happened to me or to you. One minute you're looking all spruced-up in your Sunday-go-to-meetin' dress and the next minute you're parading down the aisle of the First Baptist Congregational Church showing the parishioners your unmentionables."

I opened the cabinet door and searched for a clean rag to dry the glasses and plates. "Hence the reason I should always wear clean undies," I recapped.

"Or," she laughed out loud, "none at all."

"Nadia!" I shook my head in mock disgust.

For a long moment she cackled like an old hen then suddenly a silence filled the room. "Although, in Mrs. Pruitt's defense, I must say she is an excellent cook. Much better than I am."

"Ah, the quintessential Southern lady—always passing praise on to others." I turned to face her and noticed she had pulled a handkerchief from the pocket of her threadbare housedress and was twisting it in her hand. She was staring vacantly at a nonexistent spot on the wall.

"Are you okay, Nadia?"

My voice snapped her out of her reverie.

"Yeah." She added a laconic nod. "Have I ever told you about livin' in Benwood?"

"Benwood? Hmm… no, not that I recall." I placed the last clean glass in the dish rack. "Why don't we mosey out to the porch and you can tell me about your adventures while living in Benwood?"

A dry laugh escaped her lips. "Adventures," she repeated, "it is a tale to be told, that's for sure."

Tragedy Strikes

C hilly and damp, with thick black clouds billowing overhead, Nadia slid in the back door of the Cooey-Bentz Building in Benwood, West Virginia, where she worked as a clerk selling dry goods to the folks living in and around town. She hung her secondhand tweed coat on a hook, shook the rain from her hair and frowned when she noticed her shoes were drenched and that mud had splashed all the way up to her knees. She glanced around the store and was pleasantly surprised to find that Wanda, her coworker, had left everything clean and tidy before she had closed up on Saturday night.

Nadia had promised Mr. Bentz that she would come in early to inventory the textiles and toiletries and get requisitions ready for the following week, and she was extremely grateful he hadn't arrived to see her sporting mud-caked shoes. This job, earning forty-two cents an hour, was important for her and her brother William, and she always took extra care to appear clean, neat and professionally dressed for work. She spent hours every Saturday ironing her handmade dresses in order to give the impression that she was as elegant as the customers who frequented this upscale establishment. She also realized there were many young women in town who would take her job in a second if they had the chance, and she sure enough wasn't going to give them the opportunity.

She heard the Herschede Revere Grandfather Clock chime seven times from the corner of the shop, which confirmed that she had an hour before she needed to open the doors for customers. Nadia walked past Mr. Bentz's office and opened the closet door where the cleaning supplies were stored and scrunched down to locate an old rag or scrap of discarded cloth to scrub the mud off her legs and shoes. She slid her right shoe off and ran a dry cloth across the heel. Spitting on the cloth, she rubbed the shoe furiously. *I hate mud. Someday I'm gonna move out of Flanner Hollow, with its red clay sludge and deep ruts, and I ain't never gonna go back to that filthy place. Why can't anything be easy? Why'd I have to be born poor?* Nadia was firing herself up for a full-blown pity party, but she didn't get the chance. In a blink of an eye her world changed forever.

A blast echoed across the mountains and down through the valley. Nadia felt the building quake beneath her feet as vibrations shot unwaveringly through her body. *An earthquake? What's going on?* She pressed her hand against her forehead. *Are the perfume bottles jingling in the display case?* She glanced around in a state of confusion. Disoriented and holding one shoe in her hand, she hobbled to the window and frantically gazed out at the corner of Jacob and 36th Streets. She didn't see anything or anyone—only hazy, smoky rain.

Another explosion caused the ground to tremble in response to the falling rock and debris and the realization hit Nadia like a blow to her gut. *Benwood Mine. Oh, God! Please, no! Two blasts?* That's when she heard bedlam surfacing in the street. She looked out the window again and saw women and children hysterically screaming and sobbing as they rushed to the site.

"Pomóż nam!" a woman was screeching repeatedly as she tugged two small children behind her. Nadia didn't understand the words of the immigrants' families as they passed on the street outside the window. It didn't matter, because there were no coherent words for this.

People had migrated from all over the world to come to West Virginia to find work in the mines—Serbs, Greeks, Slovakians and Montenegrins, and of course local folks, too. She slid on her muddy shoe, ran to the door, snatched her coat off the hook, slid it over her brown cotton dress and joined the progression of panicking neighbors who were rushing toward the mine in the deluge of pouring rain. As the mine came within sight she paused to watch the black coal dust forming a plume as it forced its way through the torrential downpour. The sight caused her heart to jump up into her throat.

She inched in closer, weaving her way through the bystanders, her body trembling erratically. *The roof has collapsed. Oh, no!* Her hand rose to cover her mouth. She could see mounds of rocks and fallen debris blocking the entrance of the mine. *How are they going to get in there to save them? Is William working today?* She couldn't remember. She felt numb—totally dissociated. It was like being in a nightmare where you keep trying to wake yourself up but the Sandman won't allow it.

William had left the house an hour or so before she trudged up the muddy hollow to work, but they hadn't spoken. He was walking out the door when she woke up. *Was he heading down to Moundsville? Think, Nadia, think!* He mentioned going to Moundsville with their cousin, Mike, but she couldn't recall if the trip was planned for today. *Or did he say Tuesday?*

She edged her way closer and closer through the chaos that loomed around her. Hysterical sobs and prayers in many tongues rang in her ears.

"Tell us who's in there!" she heard a woman demand to no one in particular.

Nadia reached out and clutched a man's shoulder. "Where's the tag board?"

He shrugged.

Nadia knew that coalminers who were hired on as handloaders were assigned a place in the mine, and they would sometimes share it with a friend—or perhaps a relative. The two loaders would remove the broken coal to mine cars. Each miner was issued tags to identify a filled coal car as his work so that he would be paid based on the weight. She figured if she could find the tag board she could look at it and see if her brother was trapped inside.

"Back away!" A man covered in black soot shouted at the crowd. He shoved Nadia's slender shoulder, as she stood dumbstruck, her eyes anxiously searched for the tag board.

She pushed him back. "Where's the tag board?" she asked through clenched teeth.

"It doesn't matter!" The exasperated expression on his face did not go unnoticed. "Ya know the miners take anyone they wanna into the mines. They swap with their kinfolk to get the day off, or bring in their youngins to help 'em." His voice rose in volume, "There's no way we're gonna find out who is down there!"

Nadia knew it was true. William had done this very thing before to help out a neighbor and the company was notorious for turning a blind eye to most restrictions.

"Ye get outta here! 'Tis turning into an old-fashioned donnybrook," a man screamed with a thick Irish brogue. "We need to get those lads out of the pit! And yer gettin' in our way." His arms waved forcefully in the air as he tried to shoo the women away from the mine. A dozen or so men began constructing a barricade to keep the wives from finding a way into the mine.

Nadia felt someone gently grasp her slender elbow. She turned to find Mr. Bentz with his bifocals on the edge of his hooked nose and his shiny head low. He raised his eyes, as if to ask, are you okay? "Nadia, come back to the store with me. You can't do anything here."

"I'm sorry I left the store, Mr. Bentz. I heard the blast and ran down here." She glanced back over her shoulder toward the pit of the mine. "I don't know if William is in there or not."

"It's all right." He pushed his wire-rimmed spectacles up to his brows and studied her guardedly. "We're going to close the store down today, but we need to gather supplies and of course..." he lowered his voice to a whisper, "make some room… just in case the Cooey-Bentz Building is needed as a temporary morgue."

"No!" Nadia shook the image from her mind. "There are corners in the underground passages and the men might be huddled in them safely."

"Perhaps." Mr. Bentz's solemn eyes misted with tears that did not fall. "Hope springs eternal," he murmured, though she could perceive the trepidation he was trying so hard to mask. "Come on. Let's go back to the store and start getting things ready. We can do more good there than standing in front of this mine."

"William will be okay, I'm sure…" Nadia's voice trailed off. The notion crossed her mind that those were fuzzy and vague thoughts, like trust and hope.

"Come with me, dear," the older man said softly. "There's going to be some things happening here today that you don't need to see. Things that you'll never be able to un-see."

Nadia nodded incoherently and turned to steal another glance at the crumpled shaft before following Mr. Bentz up the blackened, dust-covered knoll.

Charleston, West Virginia

1982

"I'm so sorry, Nadia." I couldn't fathom the terror. "Was William trapped in the mine?" I waited with bated breath for her answer.

"I'm gettin' around to that part," she assured.

"I bet your parents were worried," I assumed.

Nadia twisted around on the long wooden bench, cocked her head to the side, and studied me curiously. "Haven't I told you about my parents, Dee?"

"No, you've never mentioned them."

She flicked her hand in the air. "Well, that's probably because I didn't have 'em around very long. My mama ran off with another man shortly after William was born, then Daddy died when I was fourteen."

I gasped. "That's horrible!"

Nadia gave a halfhearted shrug. "Looking back on it, I really can't blame my mama. She was raised on ham and eggs and Daddy was raised on fire bread and soppin' gravy. We were living in hateful poverty at the head of a muddy holler and she most likely wanted more out of life."

"So, you forgave her."

"I did, because holding a grudge only hurts the person hanging on to it. I had to forgive her so I wouldn't hurt anymore. Forgiveness is difficult but worth the effort."

I suppose. "What about your daddy? How did he die?"

Nadia cocked her head to the side and seemed to reflect on this question for a long moment. "He died of the Spanish Flu. He got sick and died the next day. It was unbelievably fast."

"Wow, I'm so sorry."

"He was a wonderful man whose tender heart was stitched together with steel wire. He experienced a great deal of heartache in his short life." Nadia gave me a sad smile and tucked a wisp of gray hair behind her ear.

"So, you raised your little brother?"

"Yep."

"All by yourself?"

"Yep. I quit school and Mr. Bentz offered me a job. He was a good friend of Daddy's so he helped us out, because he knew us since we were knee-high to a grasshopper."

"He must have been a very kind man."

"He was, indeed."

I took in a deep breath, trying to imagine what it would be like to lose your parents at such a young age, and suddenly noticed that the blooms on the vines threading through the trellis overhead gave off a lightly perfumed scent. "It smells great on this porch, Nadia, and these blooms are gorgeous."

She beamed. "I love this trellis. Isaac made it for me several years ago."

My attention turned to the golden branches of the Forsythia bush in front of her bay window and I noticed an empty birdcage hanging from the rafter. My gaze scanned the yard. It looked extremely tidy considering she was taking care of this place by herself since Isaac had passed away.

I knew he had died in his sleep two years earlier and Nadia had told me a few stories about him. She had conveyed that when she was barely eighteen, she married the love of her life and proceeded to bear him two healthy children in a span of four years. I wasn't overly impressed with either of her children though, because neither of them came around to spend a little time with her. *When did showing consideration to one's parents disappear in our society?* Nadia picked up a magazine from the

20

small side-table and started fanning herself. My thoughts were interrupted when she whispered, "Don't turn around but there's a handsome man walking this way."

Naturally, I turned around to look.

"I said not to turn around," Nadia hissed.

I saw Jack Kerns sauntering toward the porch with his fawn-colored boxer dog named Max.

"Jack!" Nadia beckoned with one hand. "Please, come and join us."

Mr. Kerns had been a fighter in his early years—a pugilist according to him, but a few years back he had opened a store called *Jack's Pawn Shop* that was only open on Thursdays, Fridays and Saturdays. The shop carried miscellaneous items he picked up at yard sales, estate sales and, of course, items from around town that people brought in to hock. There was always something of interest to be found at *Jack's*.

He was a stout man with scarred, gnarled hands, a prominent scar on his cheek and a slightly crooked nose, no doubt a result of his career choice, and he possessed a terrific sense of humor. I couldn't guess his age, because he most always wore comfortable shorts and loose fitting sandals, but I figured he was about the same age as Nadia. What I did know was that he had taken a shine to Nadia. I could see it in his blue eyes that sparkled with warmth.

"Mr. Kerns," I joked, "I heard all about the brouhaha at the First Baptist Congregational Church a few Sundays ago."

"Brouhaha?" he repeated, directing Max to be seated on the floor with a simple point of his finger.

"Yes," I arched my brow teasingly, "involving Mrs. Pruitt."

He hung his head and shook it slowly. "Lordy, poor Mrs. Pruitt is as clumsy as an ox." His smile exposed a gold-rimmed tooth.

Nadia tsk-tsked. "Now be nice, Jack."

I gave Nadia a teasing wink, remembering that she had told me earlier that Mrs. Pruitt did not possess the grace of a gazelle. "Nadia said Mrs. Pruitt's undies were squeaky-clean."

Nadia silently shushed me with a nudge of her toe.

"I wouldn't know 'cause I shut my eyes." Jack cleared his throat. "She's the kind of woman that looks better from a distance." He chuckled under his breath. "So, how are you ladies doin' on this beautiful afternoon?"

"We're doin' fine, Jack." Nadia motioned toward a rocking chair in the corner of her porch. "Have a seat and join us for spell."

"Thank ya, kindly." He slid the chair forward a few inches and plopped down into it. The old rocker moaned in response. He opened a paper bag and slid out a large yellow box. "I brought you some chocolates, Nadia."

"Chocolates? I love chocolate."

"I know." Jack smiled.

"My friend, Mrs. Carrington, once told me that the beauty of life is in the details. Thank you for noticing my details, Jack."

He shrugged. "I overheard you mention it one time."

Nadia beamed before mischievously continuing, "Now Jack, I told ya I was tryin' to watch my weight. When I got married I weighed ninety pounds soaking wet. I was a puny little thing back then and now, well, let's just say I've been tryin' really hard to eat a lot of salad." Nadia crossed her legs and tented her fingers on her knee. I noticed she glanced down at her well-worn housedress. I knew she owned better but deliberately chose her comfy clothes to wear around the house or when we would dart to the market, but now she seemed slightly embarrassed to be donning the old snap-up sheath.

"Nadia," Jack addressed her, "you don't need to be worrying about your weight. You are the perfect specimen of a woman." He grinned. "Besides, chocolate counts as a salad."

"Oh, hush up." She giggled and leaned toward him. "Do tell me, how do you figure that chocolate counts as a salad?"

He handed over the biggest box of Whitman's Sampler I had ever seen before presenting his illogical explanation, "Chocolate comes from cocoa, which comes from a cacao tree. That makes chocolate a plant. Therefore, chocolate counts as a salad."

My brow furrowed as I attempted to understand his reasoning. *Seriously? Is he wooing her?*

Nadia clapped her hands like a young girl with a special treat. "It makes perfect sense to me. I like the way your mind works, Jack." She tore the plastic wrap from the box and held it out to me. I like all chocolate so I plucked out the first one my fingers landed on. Jack, however, examined the contents carefully before he selected a caramel, popped it into his mouth and confessed, "I love caramel, but it sticks to my dentures, so I'm going to let it linger here on my tongue."

Nadia shrugged and yanked out a sweet treat. "Do ya'll wanna come inside? I'll fix some hot tea or coffee."

"I'd love a cup of coffee," Jack acknowledged, "I was out of my instant coffee this morning so I haven't had a cup today."

"You should've come over here." Nadia guaranteed, "I always have coffee in my cabinet. But it's real coffee not that processed stuff."

The sound of the chair's squeaking runners on the bare pine planks halted and Jack boosted himself up. "Stay," he told his friend, Max.

"Oh, he can come in too."

Jack, Max and I followed Nadia through the living room and into her undersized kitchen. We settled around the table and Max dropped to the floor and positioned his square muzzle on Jack's left shoe, while Nadia rinsed the filter basket, filled the carafe with water and opened the cabinet door. She pulled a can of Folgers from the shelf and pried the lid open. "Ya'll ain't gonna believe this." She turned to face us. "I'm out of coffee too." Nadia tilted the can forward as if to prove that she was indeed telling us the truth.

"Tea it is!" Jack declared.

"Hot or cold?"

"Either is fine with me."

I agreed with a nod of my head.

"I'll boil some water. Hot tea goes great with chocolate,"

Nadia suggested.

Jack chuckled. "But cold tea goes better with salad."

"Are you teasing me, Jack Kerns?" Nadia joked.

"Yep."

Nadia placed the teapot on the stove, ignited the gas burner and situated three mugs and a bowl full of sugar on the tabletop.

I turned my attention toward Jack. "Mr. Kerns, Nadia was telling me about living in Benwood before you arrived."

"Dee, please call me Jack. It makes me feel younger, not to mention that my dad was Mr. Kerns."

"Okay, Jack."

"So," Jack turned his attention to Nadia, "can I listen in on your tales about Benwood?"

"I don't mind at all," she assured with a nonchalant toss of her hand. She then tendered steaming mugs of Earl Grey tea along with the delectable box of Whitman's Samplers right smack-dab in the middle of the table. "Let me tell ya'll what happened next."

A Deadly Cocktail

Benwood, West Virginia
April 28, 1924

"Nadia," Mr. Bentz said consolingly, "we need to clear out the extra room so they can bring the bodies here."
Nadia's shoulders heaved up and down as she tried to control the tears that burned her eyes; a sob got wrenched loose from deep inside her and filled her windpipe. She covered her face with her hands.

"Dear, you have to try to stop crying. We'll know soon if William is alive and there's nothing you can do down at the mine. There is a rescue mission underway and there will be people coming from all around to help."

Nadia slowly raised her head from her hands, hot tears streaming down her cheeks, and she attempted to speak, but the words wouldn't come out. Mr. Bentz poured her a glass of water, handed her his handkerchief and pushed his spectacles up on his nose.

"What did you mean when you said there are things I won't want to see back at the mine?" she finally was able to whisper.

Mr. Bentz grasped her hands in his own. "Nadia," he paused momentarily, "there are going to be bodies coming out of the mine soon and many of them will be brought here. I don't know if it'd be best if you go on home or stay here."

"I'm stayin' here. I'm not leaving 'til I get word from William."

The elderly man nodded his head. "I understand but you can't go near the mine. There's most likely afterdamp seeping out the shaft and…"

Nadia interrupted him, "What do ya mean by afterdamp?"

Mr. Bentz pointedly explained, "Afterdamp is a deadly cocktail of carbon monoxide and other toxic gases caused by the fire."

She had heard the term before but at this very moment she couldn't think straight.

"Then there's the fact that… Nadia, I hate to say this but you're going to need to be mentally prepared when they start bringing the men in. I called my wife and she's walking down to help out, too. You stick with her over the next few days and you can stay at our house if you would like. We'll tidy up Allison's old room for you."

Nadia slumped by the window staring at the corner of Jacob and 36th Streets. By ten o'clock in the morning people were everywhere despite the drenching rain. A continual flow of streetcars and automobiles was streaming through town and on to the paths leading to the pit-mouth of Benwood Mine in the mill yard of the Wheeling Steel Corporation. Every now and then someone would stop in and deliver news:

> "They deputized employees to stand guard at all entrances to prevent people from gaining access."

> "There are a lot of folks banked on the hillside."

> "The B&O railroad is roping off its tracks 'cause there are so many folks walking and waiting by 'em they're concerned someone's gonna get hit by a train."

> "Some of the wives are frantically running around the isolated areas of the mill property trying to get inside."

"The Hitchman Company's crew is down there now trying to help the Benwood mill workers with the rescue."

"There are crowds of onlookers and news reporters standing behind the barricades. I ain't never seen nothin' like it."

From her vantage point, Nadia watched as The Wheeling Steel Corporation's crew, followed by a rescue truck, rolled through town, and she noticed several women, haggard of expression and wringing their hands piteously, aimlessly traversing the street outside the window.

"Mr. Bentz," Nadia shouted, "I know you've advised against it but I'm going down to the mine." Not waiting for his response, she slid on her damp coat and trudged discouragingly to join her neighbors standing vigil over the disaster.

When she arrived back at the mine, she learned that the rescue parties had shifted their focus to the airshaft at Brown's Run because the fallen slate at the main entrance was too overwhelming a task to undertake when time was of the essence. The men were taking shifts of one hour each. Wearing gas helmets, they would enter the mine and work back by degrees until a depth of about two thousand, five hundred feet was reached. Slowly, they pushed their way through the blackness and step-by-step made their way toward the main entry a mile away.

Four times they stopped during the slow journey where lifeless forms of stricken miners blocked their way. All four of them were dead. Two were badly burned; the condition of their bodies bearing unspeakable evidence of the force of the deadly blast that had swept through the mine. These four bodies were tendered back to the foot of the airshaft by eight of the forty rescuers while the others pushed ahead until the main shaft was reached. The temperature was rising quickly and the oxygen tanks grew low forcing the men to return above ground for rest and to replenish the oxygen.

Nadia realized she would be of no help here—not yet. She would come back and help in the Red Cross tent once it was set up. *Mr. Bentz had been right. I'll go back to the Cooey-Bentz Building to see if any word from William has come. No, I should go back to the cabin and see if he's home.* Nadia ran, stopping a few times to catch her breath, to the mouth of Flanner Holler.

The rain pounded on her from above and the red clay mud splashed up onto her legs from below. Nadia staggered up the steep mountain path, took a shortcut through Maslow's field and pushed open the door of the three-room shack. It was deafeningly quiet. "William!" she cried out. There was no answer. She searched through a drawer in the kitchen and found a scrap piece of paper.

William,
When you get home come to the Cooey-Bentz Building. I'll be there or at the Bentzes' house. The mine exploded. I need to know you are safe.

She placed the notelet on the table, pulled a worn wool blanket from her bed and hung it over her shoulders before trekking back out of the holler toward town. There, the hours passed slowly. News kept trickling in from the folks living in town and dozens of people were standing around, solemnly drinking coffee and waiting to help when the time came for the miner's bodies to be transported from Benwood Mine to the store.

At six o'clock that evening, Nadia accompanied Mrs. Bentz to the Red Cross tent and helped make coffee and bologna sandwiches for the rescue workers. An official statement came for the Wheeling Steel Corporation through General Manager Carpenter at dusk. "I have been authorized by President I. M. Scot to inform you of the following information."

The news reporters pushed their way through the crowd of onlookers and anxiously took notes and photographs.

Mr. Carpenter continued, "Most of the men inside the mine

entered at approximately 6:30 a.m. Then at 7:05 this morning an explosion occurred. Federal and state rescue crews are on the scene and are doing wonderful work. We are thankful for the volunteers who are also helping. As far as can be ascertained there are one hundred eleven men trapped in the mine. We are hopeful that we will be able to give a definite statement on the number of men confined inside by tomorrow."

"What caused the explosion?" a reporter asked.

"The probable cause of the explosion will be made after a thorough investigation is completed."

"Have you found any of the men?"

"Yes sir, we have. There will be no bodies removed from the mine at present. We have obtained a blueprint map of the mine workings and it will be marked, as bodies are uncovered. This will aid in the identification of the bodies."

"Is it possible that some men are still alive?"

"We are optimistic," he replied.

The rescue workers slowly picked along, working two teams at a time, until shortly after seven o'clock and when darkness fell they had reached a point only a few feet from the entrance. Reports coming by telephone from the bottom of the shaft indicated eleven bodies had been recovered. It appeared that most of the men had been incapacitated at the spot where they worked, stunned by the horrendous explosion and then suffocated by the afterdamp. The doomed men had wrapped articles of clothing around their heads in failed attempts to block the deadly gas.

"There's little hope I fear," Nadia overheard Captain Andy Wilson report as the rescue team from LaBelle reached the top of the long winding stairway leading up the shaft and stumbled into the Red Cross tent. Her heart sank. She could see that the men of the rescue team were worn, badly bruised and perspiring profusely from the problematic trip through the darkness. They were covered from head to toe with black soot. Only the whites of their eyes were discernible.

Women huddled around the gate, crying and begging in

futile efforts to get word of a husband, son—or brother, as in Nadia's case. The women, children and older folks had obviously bolted from their homes when they heard the deafening roar as the explosion occurred. They had been standing outside the mine in the pouring rain for almost twelve hours now and their blank faces were caked with salty tears. They stared hopelessly waiting for news. Their hair was uncombed and their clothes disheveled. They were exhausted but they believed with all their heart that there were men who had barricaded themselves into old workings and escaped the deadly afterdamp.

Surely, some men would survive. There had been rumblings that two men were found alive, yet this had not been confirmed and no bodies had been removed. It would be days before many of these women would learn the fate of their loved ones who went forth in the early morning hours to earn money for food by the sweat of their brows.

Darkness fell and the rain ceased. Soft cries of mourning doves intermingled with prayers being sent up to God, and hundreds of stars winked despondently in the midnight blue sky above them.

Charleston, West Virginia

1982

Nadia swallowed a lump of emotion that had formed in her throat. "I always get teary-eyed when I think back to that week. It was simply one disaster after another. It seemed as though the whole world had fallen apart."

I reached over and patted her hand. "Let's talk about it some more later. It's a beautiful day and it's getting late." I shoved the box of chocolates across the table. "Not to mention that I am stuffed. I can't eat another piece of candy." I wiped my mouth with a napkin.

Jack eyed me skeptically, "Not even the last Cashew Cluster?"

I glanced longingly at the large yellow box. "Well, if you insist." I motioned with my hand. "Slide it back over to me."

He tapped the box energetically and it glided in my direction. Turning his attention to Nadia he asked, "Did William live?"

"I haven't gotten to that part yet," she replied.

"That's the same thing she told me," I informed Jack.

Nadia shrugged. "I just wanna make sure I'm tellin' the story by following the sequence of events."

"How can you remember the details after all these years?" Jack queried, gulping the last drop of tea down his throat.

She tapped her forehead lightly. "It's etched up there. I'll never forget it. Plus, I've taken care to save every photograph I have from 1924."

I started nodding. "It is an interesting piece of history. I think I'll bring my notepad over and start taking notes. Maybe I could write it down one day so people will remember it."

Jack proposed, "Folks don't understand how quickly incidents like this can be forgotten. The whole world is moving so fast and people need to know about the past. Wasn't it Edmund Burke who wrote, 'Those who don't know history are doomed to forget it?'"

I looked at Nadia, then back toward Jack. "I have no idea."

He winked. "I think it was."

"More tea?" Nadia stood and walked over the stove.

"None for me."

"Me either," Jack agreed, "but I would like to hear more about Benwood someday."

"I'm going to come back tomorrow with a notepad and find out what happened next," I told Jack.

"Can I join you?"

I pointed to Nadia. "That's up to her."

"Oh," she huffed modestly, "I've led a less than eventful life and I've most likely talked your ear off, Jack. I'm sorry," Nadia half apologized.

"No," Jack tugged on his ear lobe, "it's still here."

Nadia and I laughed out loud at his quick, witty response.

"Did I ever tell ya'll about my first wife, Camilla?"

This was the first time I had ever talked with Jack other than for a few minutes when I dropped in *Jack's Pawn Shop* and rummaged through his two-dollar bin. "I'm sure you haven't told me."

Nadia shook her head indicating she hadn't heard any tales about Camilla.

"Well," he scratched his chin, "when we were married I used to tease her and call her 'Mother of Two' because we had two children. She didn't much like it when I called her that and one evening when we were getting ready to leave a church gathering, I said, 'Let's go home, Mother of Two.' She snickered and replied in a loud voice, 'Okay, Father of One.'"

A lopsided grin crisscrossed his face and Nadia and I cackled aloud.

"She sure got your goat, Jack," Nadia sputtered between giggles.

"Yeah, she sure enough did," he admitted. Jack stood, stretched and rubbed his thighs, "I need to go out to the market and pick up some coffee. Is there anything you want Nadia? You can ride with me or I can pick it up for you."

"There are a few things I should pick up at the store. I'll just change into something respectable."

"You look fine to me," Jack responded.

"Oh, I wouldn't wear this old thing out in public," she replied, motioning with one hand at her frayed housedress.

You wear it all the time when we go out. I considered. *Plus, I usually take you to the grocery store.* With a mental slap, I realized I was being ridiculous. *Why should I care if Jack takes her the store? Have I lost my mind?*

"I'll come and pick you up in about fifteen minutes if that works for you," Jack proposed.

"I'll be ready," Nadia purred.

With a snap of Jack's fingers, Max was by his owner's side and the two of them moseyed to the front door. As soon as we heard it close, Nadia turned to me and said, "I guess I'm gonna cancel my dinner plans with Sylvester Stallone and go to the store with Jack. He seems like he'd be a hot date, Dee. See ya later."

"So," I scoffed, "you're kicking me out."

"Mmm, hmm," Nadia cackled giddily, "and don't be crotchety!" She glanced over her shoulder. "I need to put on my pearls and apply some lipstick."

"To go to the grocery store?"

"A Southern woman never goes anywhere without her lipstick and pearls. Not to mention I have been feeling a little frumpy lately."

"Really?" I stifled a snicker. "I'll see you later. Right?"

"For sure." Nadia groaned, shooing me out of the room like a fly. I could hear the squeaky screen door slap shut behind me.

∞ ∞ ∞

The predawn light was dim as I gazed out the kitchen window and consciously observed the dense fog that always complements the soggy mist as it prances through the valley running alongside the Kanawha River this time of year. The rain from the preceding night had brought in, what the old-timers call a *peasouper,* and judging from the stodginess of the haze out-side I knew it was going to be a sticky, muggy, clammy day.

My feline friend, Gabby, leapt to the countertop and nudged me with her nose, and since I am very well trained I automat-ically reached for a can of Posh Banquet, thrust it into the can opener and watched as it twirled around in a circle. I forked the chunky meat into her bowl. "Here you go, Gabby." Her sad eyes examined me accusingly before she bounced to the floor and de-voured her tuna-flavored breakfast. She and I had been working on the "stay off the countertop" rule for three weeks now, but to no avail.

I poured a cup of aromatic caffeine and looked out the win-dow again. I couldn't see the oak tree standing only six feet from my window, nor could I see the fragrant geraniums in the planter just beyond the porch. I recognized there would be no reason to attempt to style my hair. It would be a useless en-deavor because my locks tended to curl up into Shirley Temple style tendrils when the air was damp. I started making mental notes of what I needed to accomplish today. *Check on Nadia and don't forget to take a pad of paper and a pencil. I have class at two o'clock. I need to get an oil change. I should probably put that off until tomorrow.*

My thoughts were disrupted when I heard someone urgently pounding on the kitchen door. I glanced at the clock. *Twenty minutes after seven. Who could that be at this time in the morning?* I am the type of person who needs a hot shower and a strong cup of coffee before I can speak in anything more than monosyllables so I prayed it wasn't anything important.

"Dee, it's Nadia! Open up!"

The trepidation in her voice startled me. "Are you okay?" I yanked the door open and ushered her inside. Dressed in her favorite frayed housecoat with a meat cleaver grasped in her right hand she breathlessly replied, "Somebody terrorized me all night long."

"Terrorized you? How? Are you hurt?" I led her to the table, pulled out a chair for her to sit on and filled a mug full of coffee while she regained her composure.

After a couple minutes, her breathing stabilized and she finally sputtered, "Someone started knocking on my windows at two o'clock this morning and continued to do so until the sun came up."

"Did you call the police?"

"No," she shook her head, "and tell 'em what? That someone is knocking on all my windows?"

"Did you look outside?"

"Of course," she glanced sideways at me, "but I didn't open the door and go outside."

"I would hope not." I drummed my fingers on the table. "Was it the same window every time?"

"No," she took a sip of coffee, "they would rap on one window, then a few minutes later I could hear 'em tapping at a different window." Nadia's shoulders slumped. "It continued for hours."

"Maybe it was Jack."

She looked at me like my cheese had done slid of my cracker. "He would use the door, Dee."

I nodded. She did have a point. "Well, let's telephone him."

"He may not be out of bed yet," said Nadia.

"He's going to be." I pulled the receiver from my chic avocado-green wall phone and dialed his number. "Jack, it's Dee.

Can you come over?"

"Are you okay?" I heard from the other end of the line.

"Yes, I'm fine, but someone was harassing Nadia last night. She's here at my house."

"I'll be there in five minutes."

In less than five minutes I heard Jack thumping on my door. He stepped over the threshold with Max in tow. Gabby was not pleased. She bounced up onto the countertop and released a threatening warning.

Jack's eyes darted toward Nadia and he rushed over, scraped a chair in tight beside her and held her hands in his own. "What happened, Nadia?"

She told him about the strange occurrences that took place the night before.

"Do you have any idea who it could be?"

"No." She pursed her lips.

"Why didn't you call me? Max and I would have been there lickety-split."

"And wake ya up at two in the morning? I don't think so."

Jack's voice became thick and weighty, "Sheriff Holmes stopped by the pawn shop this past weekend and told me there had been some reports of this happenin' around here. Probably some teenagers out to scare folks."

My gaze darted toward the kitchen window and I rushed over and tugged the curtains shut. Because sometimes, when it's dark outside, my house starts making noises and my imagination creates images for every creak and groan. I shuddered. Placing a cup of coffee on the table for Jack, I suggested, "I think I'll sleep over at your house tonight, Nadia. If they come back then the two of us will be there."

"The three of us," Jack announced, giving us both a so-there nod.

Max grumbled softly.

"Correction," Jack held up his finger, "the four of us. Max and I will sleep in the shed. That way if they show up again we can surprise them."

"A sneak punch?" I posed.

"Exactly." He pointed his finger at me with confidence.

Nadia wrung her hands nervously. "It's all just so upsetting, ya know?"

"I guarantee it," I replied. "But we can't do anything about it right now so we might as well get our mind off the subject."

A hush fell over the room.

"Yeah, you're right, Dee," Jack agreed. "I noticed that one of your tires is under-inflated so I'll check it out for you in a little while."

"I need to get an oil change tomorrow. I can have them check it at the garage."

Jack shrugged. "I don't mind doin' it for you and I'd hate to see you get a flat."

"Okay, thanks Jack."

I glanced from one concerned face to the other, wondering what I could do to help Nadia feel better.

"I'm just plumb worn out," Nadia confessed.

Jack and I nodded knowingly.

Max grunted and found his way to Gabby's water bowl and slurped a big gulp.

Gabby's eyes narrowed before she suddenly pounced toward the old dog that was ten times her size and then hissed theatrically. She continued to do so for several long seconds while the old dog stared back at her with wide innocent eyes. She swatted at him and he dodged her left hook—then her right —followed by a double left.

"Max must have been a pugilist like you were, Jack."

"Yep, he would have made a fine boxer. I named him after Max Schmeling."

The blank expression on my face did not go undetected.

"He was a famous boxer."

I nodded. "Ah, I see."

"Nadia," I cleared my throat, "why don't you tell us more about the Benwood Mine disaster?"

Jack bobbed his head in agreement. "Good idea, Dee."

"I'll fix us some breakfast." I opened the refrigerator and gaped at the stark supplies stored inside. *Eggs. Milk.* "French toast?"

A Temporary Morgue

Benwood, West Virginia
April 29, 1924

A t seven o'clock in the morning the telephone inside Mr. Bentz's office started buzzing. Nadia startled, opened her eyes and glanced around at her surroundings. She had fallen asleep in an overstuffed chair in the far corner and someone had placed a thick blanket on top of her. She stumbled to the dial desk telephone and plucked up the receiver. "Hello," she answered pinning the phone to her ear.

"Hello?"

She could distantly hear someone but the connection was poor and was overriding the voice on the other end. "Hello?" she repeated.

"Nadia?"

"William, is it you? Where are you?"

"Yes, it's William. I'm in Moundsville. I heard about the mine… I'm okay. I'm here with Mike."

Tears of relief flooded her eyes.

"We'll be home this evening," he promised.

"I love…" the phone disconnected in mid-sentence, but Nadia didn't care because she knew William was alive and hearing his voice relieved the burden in her heart.

"Mr. Bentz!" she yelled out. "It was William on the telephone and he's okay."

"Thank God," he replied and briskly walked to the office.

"Where is he?"

"Moundsville. He'll be comin' home this evening."

Mr. Bentz reached out and gave her a hug. "I'm so thankful."

"Yeah, me too," she admitted, though she did feel a twang of guilt knowing that many of her neighbors would be burying their kin.

"Listen to me, Nadia. They've found twenty-seven bodies and will be bringing some of them here. Most of them aren't identifiable. Well, they are pretty sure they have discovered George Holliday's body."

"Mr. Holliday? Oh, no. Are they sure?"

"They were able to identify the Masonic ring on his finger."

Nadia allowed this to sink in for a long moment. "They couldn't tell by looking at him who he was?"

Mr. Bentz shook his head remorsefully. "He was burned beyond recognition, Nadia. I'm telling you this because as the bodies come in we are taking them straight to the back of the building and I forbid you to come back there. I don't want you to see the men. It's going to be disturbing and very sad." He pulled his eyeglasses off, stuck them in his jacket pocket and wiped his eyes with back of his hand.

Tears welled up in Nadia eyes and involuntarily spilled over.

"Do you understand what I am telling you?" he choked out.

"Yes, sir."

"Stay out front here and take care of the store and help anyone in any way that you can. Mrs. Bentz and Wanda will be here by eight o'clock to help out, too."

Nadia went to the ladies' bathroom and startled herself when she glanced in the mirror. She had crusty drool down the side of her face and her hair was all matted. She tried to comb her fingers through it but to no avail. She groaned, and then realized the plight of the men and women down at the mine and scolded herself for being so self-centered. Splashing some cool water on her face helped, and she rushed to the small area that served as a cafeteria to brew coffee. She was one of the lucky ones. She wasn't one of those who would experience wrenching

grief. She wasn't one of the women who would need to identify the blackened, scorched body of her little brother or imagine him suffocating, attempting to wrap an article of clothing
around his face to keep the poisonous afterdamp from burning
his lungs.

Nadia heard the roar of vehicles carrying supplies on the
street outside. She rushed to the window and watched as trucks
and automobiles scurried in from all directions. She could hear
voices shouting and strained her neck to see what all the commotion was about. Apparently, some of the trucks carrying
oxygen tanks, food, blankets, and lanterns had stalled and the
goods were being transferred to horse-drawn wagons. She could
see, far in the distance, thousands of automobiles lined along
the narrow road leading from Boggs Run to the Brown Run's
entry. She knew the road was narrow and mostly made up of
fields of soft clay and wondered how they would manage to
transfer the supplies without sliding over the side of the steep
mountain.

The cars were speeding about in all directions and when
Nadia noticed a woman meandering in a state of confusion,
she pounded on the window and screamed, "Move!" But it was
too late. A speeding truck, bursting with provisions, struck her
from behind. The woman was propelled upwards some ten feet
before dropping onto the window of the truck, then like a rag
doll she dropped into a heap on the brick-lined path. The squeal
of brakes and the shrieks of bystanders resonated in the street.

Nadia rushed out the door. "Come here Mr. Bentz! Someone's
been hit by an automobile!" He followed on her heels.

Mr. Bentz and Nadia approached the scene, weaving in and
out of the crowd, and watching for vehicles still passing by
them, the drivers totally oblivious to what had just occurred.
They saw a man attempting to lift the woman into his arms. Her
body was limp but her eyes were wide-open, blood trickling
from her nose and mouth.

Nadia stared at the woman. She didn't know her name but
she knew who she was. An immigrant. A Polish woman whose

husband and young son were trapped inside the dark, toxic pit of Benwood Mine. Mr. Bentz lifted her into his arms. "Nadia, go to the Red Cross tent and find a doctor or nurse."

She scurried down the alley and the muddy path leading to the makeshift aid station. Panting by the time she made it there, she explained in a breathless voice, "There's been an accident and a woman was struck by a truck. Mr. Bentz took her to the store and she needs help."

Two women hurriedly picked up their satchels and followed Nadia back to town. By the time they arrived, Mr. Bentz sadly related, "She's dead." He lifted her onto a cot, reserved for the bodies being extracted from the mine, and covered her body and face with a sheet.

A reporter burst through the entrance. "Did someone in here see what happened out there?"

Nadia's eyes met his and she felt her cheeks flush. She turned away.

"Ma'am," the man dressed in a starched white shirt with two-tone shoes addressed her again, "do you know what happened to the lady who was hit by the truck?" He slid a pad of paper and a pencil from the pocket of his trousers.

Nadia turned to face him, tears brimming in her eyes. "She died."

"Did you witness the accident?"

Nadia was dumbstruck by this man's brashness. Tragedy had struck her hometown and he was asking her for details concerning the woman who was killed in a horrible accident. *Seriously? Her husband and perhaps children were mostly likely caught in the pit of the mine and she was wandering about in a state of bewilderment.* She swallowed deeply and blinked the tears away. "No. I did not," Nadia snapped.

"I'm sorry, ma'am. I didn't mean to upset you," he extended his hand. "My name is Isaac…"

She stared into his bright sage-like eyes before cutting him off. "I don't care what your name is, sir. Benwood is suffering here." Nadia made a sweeping motion with her hand. "This catastrophe is a misfortune that will forever affect our little town and we really don't need reporters asking questions regarding…"

Mr. Bentz sprinted across the store where Nadia was standing, visibly upset. "Excuse me. If you are a reporter you must leave." He pointed to the door. "We have enough commotion to deal with at this moment and the store is closed."

"Yes, sir. I apologize." He offered a polite nod of his head, gazed into Nadia's dark-lashed, almond-shaped hazel eyes for a brief moment then walked out the door.

"Nadia, you should go to my house and stay with Mrs. Bentz. You need to get some rest," Mr. Bentz suggested.

"I ain't leavin'. There will be people who need comfort and women coming in later to identify their husbands. I'm a part of this community and I am gonna do what I can to help." She squared her chin.

"Okay, fine." Mr. Bentz glanced around the store. "Gather all the linens and anything you can find to cover the bodies with and keep the coffee brewing." He gently grasped her arm. "Nadia, no matter what goes on around here today I insist that you do not go into the back room where they are placing the men. Do you understand me?"

Nadia nodded. "I do. I'll stay out front here and take care of the store."

"And do not, under any circumstance, allow anyone to enter the back room through this door. The rescue workers are using the back entrance and we don't need hysterical women rushing in there."

"But won't you need 'em to identify their kinfolk?"

Mr. Bentz's voice lowered to a whisper. "Nadia, dear, most of the bodies will never be identified. They were seared by the flames which swept through the entries or crushed beneath the fallen stones."

One-by-one dead bodies were carried out of the mine and at the end of the day the improvised morgue in the Cooey-Bentz Building was filled to capacity with motionless corpses.

"They are setting up more temporary morgues in the fields surrounding the air shaft at Browns Run." Nadia overheard Mr. Bentz telling the vicar later that evening. "We must make arrangements for the burials immediately."

"I know," the tired looking priest ran his hand through his hair. "The odor of decaying bodies will be overwhelming very soon."

Nadia's stomach heaved, and then knotted. Afraid to open her mouth for fear she'd throw up, she ran to the ladies' bathroom.

Charleston, West Virginia

1982

By the time I had whisked the milk and eggs together, sprinkled the cinnamon and sugar on the tops of each piece of bread in the skillet, and cooked them until the undersides were brown and crisp, Nadia had finished telling us another portion of her tale. I piled two plates with cinnamon swirled French toast and placed them in front of my guests. "Syrup?" I asked.

"Nah," Jack replied, "a dab of butter will be fine for me."

"Butter is very nutritional," Nadia inserted, shaking her fork ever so slightly in the air. "And lard is good for ya too. It's chock-full of vitamins."

Really? Since when? I refilled their mugs and placed the maple syrup on the table.

"You're not eating, Dee?" Nadia asked, as she smothered her French toast with the syrup.

"No, I don't generally eat this early in the morning, but you two dig in."

I pulled up a chair, plopped down and thoughtfully sipped on the healthy dose of needed caffeine. "So Nadia, was the driver of the truck prosecuted for hitting the woman on the street?"

"No. I later found out that she had just found out that her husband and her only son were killed in the mine and she was aimlessly staggering around town. She was so grief stricken that she didn't know where she was or where she was goin'. It was sad." She lifted a fork full of toast to her mouth. "It really wasn't the driver's fault."

Nadia started to splutter, then cough.

"Are you okay?" I asked.

She dipped her chin and held up one finger indicating she'd be all right. "Goodness gracious," she whispered, scrunching her nose as Jack began to choke and pound his chest.

"Are you okay, Jack?" My attention turned to him.

He nodded. "It went down the wrong way."

"Does the French toast taste all right?"

They both started nodding energetically. I noticed Nadia cut off a minuscule bite and force it into her mouth.

"Are you sure?"

"Mmm, hmm," Nadia said.

Jack nodded again.

I noticed him slip a large hunk of his breakfast beneath the table for Max and saw Gabby scurrying over to see what treat the dog had been given.

"De-light-ful," Nadia said, carefully over-pronouncing each syllable while flashing a knowing look at Jack.

"Indeed," Jack cleared his throat.

They were both nodding so vehemently that I imagined their heads might fall off. I briefly wondered if they had suddenly developed some rare, dreaded nodding disease, as my eyes darted from Nadia to Jack and back to Nadia again. I heard Max heave before spewing out a slimy blob on my right foot. I lifted the tablecloth and looked at the floor. Sure enough, Max had upchucked and my bare foot was now covered with half-chewed French toast. *Geez.* I shook my foot a few times and reached for a napkin, noticing Gabby tripping over to staunchly position herself between Max and my foot. *You're a little too late, Gabby.*

"Okay, what's up?" I pointed to their plates. "Is it horrible?"

"It's fine," Nadia cleared her throat. "Did you try a new recipe out?"

"No. It's the usual bread, milk, sugar and cinnamon recipe," I shrugged my shoulders.

"Are you *sure* you used sugar?" Jack inquired.

Nadia arched an inquisitive brow echoing his question.

Bolting up in my chair, I suddenly recalled the time I made sweet tea for the family reunion, only to discover later that I had snatched the Tupperware container that held *salt* instead of the one that stored *sugar* and generously added a heaping cup full to the steeping pitcher of tea.

"I am *so* sorry." I pointed at the perfectly browned bread. "Can I taste it?"

They both shoved their plates in my direction. I pinched off a smidgen and reluctantly placed it on my tongue. I grimaced. "Yuck!" I spat the bite into a napkin. "No wonder Max threw it up." I picked up their plates, scrapped the remains of breakfast into the trashcan and dropped the dishes into the sink. Turning, I propped my hand on my hip, and in a terse voice informed them, "Look guys, if I serve you something that tastes bad you can tell me. I mean if I accidentally put salt instead of sugar in a dish I won't be offended if you just tell me. Okay?"

They nodded. Again.

"Look Dee," Jack said, "I've eaten more dangerous food than that French toast of yours and lived to tell about it."

Dangerous? I've never had anyone refer to my cooking as dangerous.

Nadia tilted her head inquiringly. "What kind of dangerous food have you eaten, Jack?"

"The most dangerous food in the world, of course—weddin' cake."

They both started laughing like two youngsters pitchin' woo, while I scrubbed my foot with a paper towel and wiped up the mush from underneath the table. "Can I fix you both something else?"

"No," Jack shook his head convincingly. "I need to get down to the pawn shop and sort through some things I bought at an estate sale."

Nadia fervently waggled her head in tune with Jack's. "I'm really not very hungry."

∞∞∞∞

I arrived at Nadia's at exactly seven o'clock that evening. I rapped on the door twice before Jack slung it open. "Where have you been?" He tugged on his belt, pulling his trousers more firmly about his hips. "Nadia and I have supper about ready."

"Great, I'm starved." I could smell her famous meatloaf mingled with the scent of lemon furniture polish as the screened door flapped closed behind me.

"C'mon in, Dee," Nadia called out from the kitchen. My stomach rumbled in anticipation when I spied the dishes she had spread out on the countertop. "Fill your plate and have a seat. We're eatin' buffet style tonight."

She didn't have to ask me twice. Nadia had made one of my favorite meals in the whole wide world. There were bowls of collard greens, mashed potatoes, slow-simmered green beans wilted in bacon grease, and a pan of meatloaf scattered across her small workplace. She cut a portion of meatloaf, deposited it on a paper plate and placed it on the floor for Max.

"You don't have to give Max any food. I fed him before we came over," said Jack.

Nadia pursed her lips. "Jack, Max is also a guest in my home and I wouldn't dream of him having to smell this meatloaf without giving him a slice."

We settled around her small table and after Jack took a right long spell saying grace, we dug in. It was scrumptious. Jack talked throughout most of the meal about sorting through the miscellaneous boxes he had bought at the estate sale and how much he figured he could sell them for at his shop. "One of my favorite pieces," he said, "was a pink Cadillac cookie jar."

"That sounds like a good find," I acknowledged.

"I also picked out a couple things for you ladies," he winked at Nadia.

"Oh, Jack, you didn't need to do that."

"Well, I wanted to." He reached over and plucked a brown paper bag from the floor. Slowly, as if he were a game show host, he pulled a set of flour sack tea towels out and presented them to Nadia.

"Thank you, Jack. Flour sack tea towels are my favorite."

"I know." He nodded slowly. "And for you, Dee," his eyes twinkled brightly, "I found this." Jack placed a book on the kitchen table and slid it toward me.

I gently ran my fingers over the intact, but well-worn cover. The spine of the book showed modest wear, and the pages fell open easily. I gasped. It was an 1879 original Mark Twain book entitled, *A Tramp Abroad.* I tenderly opened the cover only to find that it was a 1st Edition. "Oh," I gasped again. "This is beautiful. I love Mark Twain."

"I know you do." Jack smiled.

"He has such great narrative authority and most of the time I feel he is cracking himself up when he writes. Have you read Mark Twain's *Burlesque Autobiography*? It's hilarious!" I didn't wait for their response and energetically rambled on before quoting Mark Twain, "I was born without teeth—and there Richard III had the advantage of me; but I was born without a humpback, likewise, and there I had the advantage of him." I giggled underneath my breath before continuing, "My parents were neither very poor nor conspicuously honest."

Jack and Nadia offered an amused half smile.

I sighed. "And then there's his *Awful, Terrible Medieval Romance,* which is a hoot!" I glanced from one expressionless face to the other. "Yeah, he's funny. Anyway, thanks Jack. I love it. It's the best gift in the world!"

"I'm glad you like it." He replied. "I haven't read much of Mark Twain's writings, but I do enjoy reading Friedrich Nietzsche."

"Really?" I was impressed. "He was a German philosopher

and his writings influenced Sigmund Freud."

"Yep." Jack dipped his chin.

"Do you have a favorite quote?"

"I do. It's 'And those who were seen dancing were thought to be insane by those who could not hear the music.'"

"Profound." I murmured.

"Yeah," he agreed.

I turned my attention to Nadia. "How about you? Do you have a favorite author or book?"

"The Bible is a favorite book and one of my beloved promises is, 'For I know the plans I have for you,' declares the Lord, 'plans to prosper you and not to harm you, plans to give you hope and a future.'" She puffed up.

"Amen," Jack witnessed her statement.

"Jeremiah 29:11," she added as an afterthought.

"The Bible is full of encouraging assurances," I confirmed, "and that verse is lovely."

"Thanks." Nadia's rosy cheeks brightened.

"The Bible is the most important book ever written," Jack said with a smile that fully exposed his gold-rimmed tooth. I could tell he was sweet on her. "Now," he bent forward and in a conspiratorial tone of voice informed us, "we need to set up our game plan for tonight." He stood and closed the window curtains. "We are gonna catch those little rascals who were trying to scare Nadia last night."

Nadia and I leaned in and confessed with confidence. "We're all in. What's the plan?"

"First," he motioned toward the living room, "we are gonna relax for a spell and have Nadia tell us some more stories about living in Benwood. Then once the sun goes down we'll secure the perimeters."

"Let me grab my memory album," Nadia stood. "I'll be right back."

The Raging River

Benwood, West Virginia
April 30, 1924

T he Ohio River was wide and running fast with murky water from the recent rains. Gusty weather came in and dark clouds sped over the hilltops trailing veils of rain. The rescue workers were still finding and removing bodies from the mine and with the odor of decaying bodies becoming overwhelming all hope was lost. Women and older folks still stood diligently outside the shaft.

Nadia slid her key in the back door of the Cooey-Bentz Building and motioned for William to go in front of her.

"Mr. Bentz? Are you here? William and I are gonna go down to the Red Cross tent today and help serve food unless ya need me here."

He didn't reply.

"Mr. Bentz?" She scanned the store and walked toward the office area. She could overhear him talking in a low but angry voice. She rapped quietly on the office door.

"Yes?" he replied without opening the door.

"Good morning, Mr. Bentz. It's Nadia and I just came to tell you that William and I are gonna go down to the…"

"Nadia!" The door swooshed open in a welcoming way. "I'm so glad to see you!" An elegant woman wearing an orange and brown, fine silk crepe dress with a graceful pleat down the front smiled at her. The woman looked hauntingly familiar.

Nadia blinked once and then twice. "Mother?" She stared at the woman for a long drawn-out moment. The woman had chopped off her long russet locks and was sporting a short, smooth bob cut.

The woman reached out and grasped Nadia's shoulders. "Just look at you. You're all grown up." She attempted to pull her closer but Nadia jerked away.

"I am grown up. It's been ten years since you've seen me, Rebecca."

"So," Nadia's mother arched a brow, offering a veil of a smile. "I'm Rebecca now?"

Nadia looked at her stonily as her arms crossed in front of her chest. "Well, you're certainly not *my* mama," she spat. "You've been gallivanting... wherever you were... and now?"

Mr. Bentz cleared his throat. "Would you like to spend some time here in my office? I have some things to tend to." He gave Nadia a sorrowful smile and lowered his head.

Rebecca's attention turned toward William. "Thank you, Mr. Bentz. I would love to spend a few moments with my children." She motioned for William to join her in the office. He turned away, his chin trembling, and then looked at Nadia as if waiting for some cue.

"Sure," Nadia growled through clenched teeth, "we have a couple minutes." She momentarily considered running from the room and sneaking down the fire escape. Instead, she reached out and grasped her brother's hand. They stepped across the threshold and she slammed the door closed behind them.

Nadia demanded morosely, "What are you doin' here, Rebecca? Huh? Did you forget something when ya ran out of town? Perhaps your children?"

"Oh, Nadia." Her face filled with confusion and something that resembled shame. "It's complicated."

"Complicated," Nadia repeated. "Yeah, well our lives have been a little complicated too." She tilted her head slightly. "Did you know that Daddy died?"

Rebecca nodded slowly. "Yes, I did hear the news." She fluttered her hand in the air. "But not until after the funeral."

Nadia smirked, realizing her fists were clasped so tightly that her nails were digging into the palms of her hands. "What do ya want?"

Rebecca sighed, "I heard about the mine disaster and I was concerned for William. I was afraid he might be working in the mine when it exploded." She stretched her arms out to him.

Through steel eyes, William flatly responded, "I wasn't."

She faltered. "I can see that."

"Is there anything else?" Nadia asked. "We were about to go help feed the folks who are standing vigil down at the mine."

"Well... well, yes. I wanted to let you know that I will be moving back to Benwood. My husband, James, has accepted a position working at the Bank of Benwood and I had hoped that, once we get settled, you both could come and visit from time to time." Rebecca's eyes pleaded for understanding.

Nadia's heart plummeted to her toes. She couldn't believe what had just tumbled out of... her mother's... this woman's mouth. She fought back the tears that were threatening to fall, swallowed the lump that had formed in her throat and offered her hand to William. With a proud lift of her nose, she demonstrated that she was a lady not to be trifled with. "We'll keep your generous offer in mind." Her eyes linked with William's and she slanted her head toward the door. "Let's go, William. There are folks out there that need us."

Mr. Bentz was cutting large, dark pieces of cloth and folding each of them into tidy squares when Nadia and William exited the room. He rushed to them. "Are you okay?"

Nadia dipped her chin. "That wasn't the mama I remember. I'm just surprised, I guess."

"I was too when she showed up here this morning. I'm sorry."

"It's not your fault," Nadia gently placed her hand on his arm.

"I know, dear. It's..." he shook his head, "it's just that I worry about you and William."

"We will be fine, Mr. Bentz."

He offered her a knowing smile and handed her a stack of cloth to carry down to the mine. "They need more material to cover the bodies," he explained.

Nadia nodded marginally, as though this was the most natural comment a person could make, and tucked the bundle under her arm. The office door behind them squeaked on its hinges as it slowly opened, and since Nadia didn't want to face Rebecca twice in one day, she tugged William toward the back door and they hurriedly escaped into a drizzling rain and gloom that engulfed Benwood.

"Why did you sugar coat it back there, Nadia? Why didn't ya tell her what you really think?"

Nadia laughed in spite of herself. "That was sugar coating it, William."

"Maybe she's changed," he suggested.

"You're making an assumption, William. Just 'cause a chicken has wings don't mean it can fly."

He pondered on this for a few minutes before asking, "Why are we goin' this way, Nadia?"

"I need to clear my head," she responded.

"So do I," he admitted. "I didn't even realize who she was. I didn't remember what Mama looked like."

Nadia opened her mouth and was about to tell her brother that she no longer considered *that woman* their mother, but quickly decided not to share her distasteful opinion of *their mama* with William. She knew it had to be a shock to him because she was certainly taken aback. She honestly didn't know what to think about her sudden reappearance. *Ten years is a long time. Did she actually think we'd be happy to see her?*

Once outside, Nadia realized it wasn't raining as much as misting, the tiny traces of water seemed to hang stagnant in the air instead of dropping. She glanced down to see a copy of *The Wheeling Intelligencer* and scanned the headline, "LAST RAY OF HOPE FOR LIVES OF MINERS GONE." She shuddered, still finding it difficult to believe that this nightmare was real.

She and William walked in companionable silence, turned at Chapline Street then continued to Water Street. She knew this would add a few minutes onto the one and a half mile walk but she desperately needed to get her thoughts together. *What will William and I do with Mama… with Rebecca back in town?* The woman's words echoed in her ears. *Come and visit from time to time? Talk about a slap in the face!* The longer she ruminated the angrier she became. *The nerve!*

Her angry reflections were cut short when she noticed a woman at the bottom of the hill. The woman was standing on the tip of her toes on a flat rock beside the Ohio River, wearing nothing but a thin dress. She looked like a bird about to take flight. Nadia tilted her head and tried to comprehend what the woman was doing standing there, so close to the overflowing banks. She recalled the day when she stood on the bridge; weeks after her father had died, and contemplated hurling into the river so that all the pain would go away. There are breaking points in life, Nadia knew, moments when you think that you'd be better off dead. She prayed this was not what the woman was thinking. Nonetheless, much to Nadia's astonishment, the woman leapt—lunging headfirst into the water.

Nadia instinctively screamed at the top of her lungs. She and William stood dumbstruck with a few other people as they watched a woman struggling against white foam and debris that was gushing downstream in the gray-green water. Then she watched in horror as a man jumped, feet first, into the raging river. *What is going on?* The entire scene was surreal.

The man fought against the hissing force of the plummeting water toward the woman, but she was caught in a current that was carrying her downstream with great force. He kept swimming toward her, occasionally pausing for a brief minute to take a deep breath, all the time fighting against the current. The crowd gasped with admiration when they saw the man had reached the drowning woman and was now attempting to pull her back to the bank of the river.

Folks scrambled down the sandy hilltop to help him lift her limp body from the dark river. Cupping a hand over her eyes, Nadia watched with interest as the unknown hero stumbled to his feet. Applause broke out around her.

He seems vaguely familiar.

They continued to watch as folks carried the woman up the hillside with her drenched rescuer in tow. When they reached a flat area, a few yards away from where Nadia and William were observing the strange occurrence that had just taken place, the sopping man caught her eyes with his own. His gaze seemed to reach inside her. He offered her a polite nod of his head before the town folk ushered him away.

She recognized those sage-like eyes. *The reporter.* She noticed the two-tone shoe he was wearing. *He must have lost one of his shoes in the river*, she reasoned. *What's his name? Ah yes, his name is Isaac.* As she watched him hobble, drenched and wearing only one shoe, she stared at his broad shoulders and her mind started to race. Her daydream caused her cheeks to flush. *He is like a tall, cool drink of water on a hot summer day.*

Charleston, West Virginia

1982

"Why did the woman jump into the Ohio River?" I asked, wrapping myself in an afghan and flinging myself down on the couch.

"She was a grieving widow and heartache can be overwhelming, especially at first. Her husband was identified as one of the deceased."

"That's sad," I commented. "I guess Isaac was quite the hero, eh?"

Nadia nodded ardently. "Bad things happen in the world, like war, illness and mining disasters. But out of those situations always arise stories of ordinary people doing extraordinary things. Let me show you a picture." She turned the pages in her album and plucked out an old black and white photograph. "Look," she pointed to a group of rescue workers wearing oxygen tanks as they entered the coal mine. "These men were heroes. They were most likely scared out of their wits, but they went in anyway."

I leaned back on her overstuffed plaid couch and stared at the ceiling, feeling the cold breeze of the window-unit air conditioner flowing across my face. "Did you hate your mother? The only reason I ask is because my mama is my rock, my sounding board and my champion. I couldn't imagine life without her."

"I was angry with her. She had abandoned us and then

showed up suddenly, acting so nonchalant, like she had simply come home late from the grocery store." She looked at me reflectively. "I don't think I really hated her, but sometimes one can mistake hate for anger." She paused transitorily. "I can assure you that I resented her for leaving us."

"I have no doubt."

"I don't think Rebecca really loved me, Daddy, or William." She shrugged dismissively. "But marriage is hard and goin' through tough times is harder still. There never is a lane so long that it don't have some hills and curves."

"That's for sure," Jack interjected. "There are some secrets to having a good marriage."

"Like what?" I asked.

Jack rubbed his chin, deep in thought. "Well, for example, a husband should always keep his willy securely zipped up in his trousers when his wife ain't around."

I laughed out loud.

"That's for sure," Nadia agreed. "Probably one of the most important things to remember. Another thing," she added pensively, "is that a husband should always look at his wife like she's the only woman in the room. My husband did that and I always appreciated it."

"Mmm, hmm," Jack concurred. "You should never give your horse more attention than your wife, unless you like sleepin' in the barn." He chuckled out loud. "And women need to realize that husbands are the best people to share their secrets with. They'll never tell anyone, because they aren't even listening."

I shook my head, wondering how long they would go on with this banter.

Jack continued, "One time my deceased wife, Wanda, and I were gettin' ready for bed. She was standing in front of this full-length mirror and looked at herself for a spell. She shook her head and muttered, 'When I look in the mirror I see an old woman. My face is all wrinkled, my hair is gray, my shoulders are hunched over, I've got fat legs, and my arms are flabby.' She turned to me and pleaded, 'Jack, please tell me something positive to make me feel better about myself.' I wasn't thinkin' clearly that night because I had just came home from a boxing match, so I murmured, 'Sweetie, there ain't nothin' wrong with your eyesight.'"

"Ouch," I cringed.

"Yeah," Jack shook his head remorsefully. "I wish I would have thought that one all the way through."

"What do you wish you would have said?" I probed.

He considered this for a long moment. "I wish I would have told her, 'We're like a couple of prunes, as time goes by, we're gettin' wrinkled, but a whole lot sweeter.'"

Seriously? I speculated on his choice. "I don't know, Jack."

"It would have been better than telling her that nothing was wrong with her eyesight," Nadia defended him.

"Perhaps." I rolled my eyes.

"Here's another truth about marriage. A woman marries a man expectin' he'll change, but he don't. A man marries a woman expectin' she won't change, but she does."

Nadia laughed out loud and I simply shook my head. "Enough advice about marriage, guys. What's the plan for tonight? How are we going to catch the interloper, or interlopers, if there's more than one of them?"

Jack stood and began pacing around the room. "Well, I brought some things up from the pawn shop. Max and I strung a cord around the house and hung a cowbell on it. This way, if they come close to the house they'll trip on the string, which will cause the cowbell to clang. Then Max and I, who will be stationed in the shed, can surprise them and either tackle them or," he shrugged, "knock them out."

Nadia's eyes widened. "Knock 'em out?" she squeaked.

"Yeah, I know how to thwack a man just hard enough to drop him to the ground but not cause too much damage." He ran his damp hands down the legs of his trousers, and nodded as if his declaration settled the matter. "I was a professional boxer," he reminded us.

"What if they have weapons?" Nadia countered.

"I brought a Taser from my pawn shop too."

Nadia's jaw dropped open. "What's a Taser?"

Jack explained, "A fella named Jack Cover, who was a NASA researcher developed the Taser device a few years ago and named it after a book featuring his childhood hero Tom Swift. It was called *Tom Swift and His Electric Rifle*. It's basically a weapon that fires barbs attached by wires to batteries and it causes temporary paralysis. It won't kill 'em."

Nadia wrung her hands nervously. "How on earth did you end up with a Taser in your pawn shop, Jack?"

Jack pressed his lips together tightly. "I don't quite recall."

"You don't recall or you won't tell us?" I asked.

He pointedly stared at me. "Does it matter, Dee?"

"You're right." I held my hand up in defeat. "It's none of my business." My eyes narrowed to slits. "Jack, are you a military man? The only reason I ask is because you seem to know quite a bit about securing perimeters and Taser guns."

"Well, yeah. I served in the Army."

"Really? What did you do in the Army?"

He shrugged. "A lot of different things, most of which I don't care to recall."

"Tell us something, Jack."

"All right, I was in the middle of hell back in 1943. My tail gunner was a fella named Glenn Harpold. He was a car mechanic from Florida and he had my back right until the day we left. When we came back to the States, Glenn and I served another year as Death Notifiers."

"Death Notifiers?" I echoed. "Is that the same thing as a Casualty Assistance Officer?"

"Yep." He sighed. "Most of the time the military would send telegrams, but when it was possible we were sent to the homes of relatives when someone was killed in action to inform them of what happened. I can assure you that no one wanted to open their doors to us."

"I have no doubt," I replied. "Sorry I asked."

"It was many years ago," he replied with a shrug. "It wasn't until the Vietnam War that initial notice by telegram was replaced by personal contact from a uniformed officer completely. Back during World War II, it depended upon where the family lived. If they were close enough to the base, Glenn and I would have to call on folks personally."

"I did that one time," Nadia said with a shudder. "I wasn't trained or bonafide to deliver the news, but it was horrible. I accompanied Mr. Bentz to inform Mrs. Piehowicz that her son had been found dead in the mine. The look in her eyes when she opened the door and saw us standing there is like a snapshot that's etched in my mind. She knew why we were there. Mr. Bentz didn't need to say a word."

"It's a hard job, it's like confirming the residents' worst fears, and I can honestly say that I recall every single time I had to do it," Jack's voice grew gravelly.

With a shake of his head, he walked over to a box he had tucked in beside the gas burner in Nadia's living room. He rummaged through it and withdrew three walkie-talkies. "We'll stay in contact tonight using these." He handed one to Nadia and then to me. "If ya'll hear anything all you need to do is push on this button and talk into there." He pointed to the device indicating the exact location.

"What are we supposed to say?" I inquired.

He crinkled his brow. "Just tell me you heard something and where you think you heard it comin' from."

"No." I brandished him with a flip of my hand. "We have to have a secret password. Or perhaps a code." It suddenly occurred to me. "The Eagle has landed. That's what we'll say."

Jack and Nadia exchanged glances.

"The Eagle has landed," Jack reiterated.

"Yes." I was pumped now. "Just like Neil Armstrong when he relayed a message to the NASA Mission Control in Houston in 1969. He simply said, 'The Eagle has landed,' and with those words, the dream of President John F. Kennedy's notion to put humans on the moon by the end of the decade became true." I nodded my head as if my story conclusively settled the matter.

"It's perfect. It will be just like we are on a mission—kinda like private investigators." I pressed my fingers to my forehead. "I love *Magnum P.I.,* and *Hill Street Blues*."

Nadia joined in. "We'll be just like Jethro when he was a double-naught spy in *The Beverly Hillbillies*."

Jack ogled us wearily.

"We'll be incognito!" I tooted, before turning my attention to the expressionless-faced man. "Those who could not hear the music, Jack," I reminded him of his favorite Friedrich Nietzsche quote.

"We've got your back, Jack," Nadia assured.

He and Max moaned in unison.

Nadia swathed herself in a cotton robe, rushed from her bedroom and shook my shoulder unmercifully. I woke to find her staring at me, her face contorted with anxiety. "Wake up, Dee. I heard a knock on my bedroom window."

I bolted up, clumsily causing my copy of Twain's, *A Tramp Abroad* to topple onto the floor. "Did you radio, Jack?"

"No." She shook her head fervently. "I couldn't figure out how to use the walkie-talkie."

"Okay." I fumbled underneath the patchwork blanket and finally located the one I had been assigned. It must have slid beneath the cushions after I had fallen asleep, cuddled contently on her sofa, because it took me a long moment to find it. I was about ready to push the "talk" button when we heard a *tap, tap, tap,* on the living room window. *The perpetrator is right outside this window.* Nadia and I gasped, held our breath and slowly sank down low to the floor. "The Eagle has landed outside the living room window," I pushed down on the switch and informed Jack in a barely audible voice.

We waited for a response. There was none.

When we heard the knock on the door we both nearly jumped to high heavens. "Oh, no!" I pulled the walkie-talkie close to my lips. "The Eagle has landed and he's knocking on the front door, Jack! Are you there? Where are you?"

Even while I was saying the words I could hear them outside the front door reverberating back at me. Nadia and I stared at one another with wide eyes. "Oh my, what if they've knocked out Jack and stole his walkie-talkie?"

Thump, thump, thump. "Open up, it's me. Jack. Let me in," we heard a man shout from the porch.

"Does that sound like Jack to you?" I asked.

"Yeah, but what if someone is holding him hostage and they're forcing him to knock on the door?"

"True," I agreed. "What should we do?"

Then we heard a voice booming through the handheld device. "It's me, Jack. I'm knocking on the door. It's all clear."

Nadia pointed to the two-way radio transceiver. "Ask him what the secret password is," she suggested.

"Right." I nodded in agreement. "What's the secret password?"

"Seriously?" We heard in response.

"Look, Buddy. We're not opening this door until you can tell us what the code is."

A grunt was followed by the top-secret words we longed to hear. "The Eagle has landed! Okay? Will you let us in?"

"Who is out there with you?" I probed, still considering the possibility that this may be a hostage situation.

"Max," Jack responded flatly. "It's Jack and Max and we're standing on your porch," he had abandoned the use of his hoity-toity radio device and was now shouting at the bolted entrance. "Please, let us in," he articulated deliberately.

Nadia and I glanced at each other and bobbed our heads in accord. She rose from her squatting position and proceeded to unlatch the deadbolt, unhook the chain and finally twist the knob to allow them to enter. Jack and Max stayed outside and motioned for us to join him. "Look ladies, we didn't see anyone out here but we did hear some tapping, so we started looking around. All we've seen is that." He pointed in the direction of the birdcage on Nadia's porch. "Just like Sherlock Holmes once said, 'Eliminate all other factors and the one which remains, must be the truth.'" We strained our necks and our eyes followed the direction Jack had indicated. I could see a small bird balanced on top of the cage.

Suddenly we heard the rich melodious song of the fella. A small European thrush with drab brownish plumage bolted through the open door, brushing slightly across the big man's shoulder. Jack's hands started whipping around totteringly when he felt a wing brush against his ear. "What in the tarnation?"

"It's Bing!" Nadia exclaimed with excitement.

"Who is Bing?" Jack and I asked at the same time.

"He's my nightingale. He's been missing for three months!" she exclaimed.

We watched as the graceful bird alighted daintily on the dog-eared antenna of her outdated television like a hummingbird taking the measure of a flower. Nadia immediately ran over and scooped him up gingerly into both hands. "Oh, Bing! I'm so glad you came home. I've been so worried about you."

Jack pointed his finger at Nadia's feathery friend. "I think he's the one who has been tapping on your window at night."

"Nightingales come out at night?" I said aloud before realizing how stupid I sounded.

"Night. Nightingale." Jack looked at me as if inquiring whether or not he should continue.

"Got it." I shot him a soppy grin.

"Jack," said Nadia, "would you fetch the birdcage that's hanging on the porch and bring it in the living room?" She pointed to a tall pedestal table in the corner of the room.

"Sure." He was back in a second and positioned it where she had asked. Max padded in behind him and dropped, uninterested, onto the floor.

Nadia placed Bing on top of the cage and opened the door for him. He hopped down and freely settled into the wiry enclosure and started crooning lyrically.

"Wow," I said. "Bing is talented."

"Yeah." She smiled proudly. "He's quite vocal at night, especially during breeding season."

"Is it breeding season for nightingales?"

"I don't think so." Nadia grasped her hands together joyfully. "He must have just missed me." She turned to face us. "I'm so excited and I know I won't be able to go back to sleep. Would you both like to join me for a cup of Chamomile tea?"

Jack accepted the offer immediately. "Sure, it sounds good."

"Will you tell us what happened next in Benwood?" I asked.

A Hero in the Midst

Benwood, West Virginia
May 1, 1924

Nadia noticed him as soon as he entered the store. It was the man who saved the drowning woman. *Isaac.* He had scrapes on top of his bruises and scratches on top of the welts. She smoothed the hem of her dress.

Since Mr. Bentz was helping coordinate the burials of the men and boys who were killed in the mining accident, she was in charge of the store for the day. So, she lifted her chin and approached him with a smile. "Can I help you?" she asked.

He chuckled under his breath. "As a matter-of-fact you can." He pointed to the floor and she noticed he was wearing an old pair of shoes that were far too large for his feet. "I lost one of my shoes yesterday and I'm wearing a pair of loaners."

She recalled seeing him, drenched after the lifesaving event in the Ohio River and had noticed that one of his shoes had been lost in the raging water. "I saw how you lost your shoe. You are a hero."

"Nah, not really. I was just doing what anyone would do." He offered her a piece of Wrigley's chewing gum.

She graciously accepted, figuring he must be rich if he had a pack of gum in his pocket. "That's not true." Nadia smiled. "You were the only one who attempted to save the poor woman." She motioned for him to follow her. "By the way, my name is Nadia."

"It's nice to meet you, Nadia. I'm Isaac."

Nadia's stomach whirled nervously. She remembered his name but didn't let on. "I am assuming you'd like to buy a new pair of shoes that actually fit your feet." She arched a brow anticipating his response.

"Yes, ma'am."

"What size?"

"Twelve."

"You're in luck. We're running short on most all supplies, since the accident at the mine, but we still have shoes in stock." She pointed to a long, narrow table that held shoes of all sizes, shapes and colors. "Does anything suit your fancy?"

He raised his gaze to meet hers and she flushed when she stared into those sage-green eyes.

"I do see something that suits my fancy," he murmured before dropping his eyes to examine the shoes. "What size are those?"

Nadia plucked up the pair he indicated and examined the price tag and size tucked inside of the pair. "These are a size twelve and so are these." She pointed to the four pair of shoes that would fit him properly.

"I'll try those on." He sat on a bench and slipped the plain, black dress shoes onto his feet. "These will work fine, ma'am. I'll take them. No need to package them up, I'll be wearing them out of the store."

"Would ya like a paper poke for the loaners?" She pointed to the well-worn shoes that were now in his hand. "Yes, thank you. I will need to return these."

"Perfect, then let's get you checked out."

He followed her to the cash register where she punched on some keys. "Two dollars and fifty-five cents." She pulled on the hand crank that opened the cash drawer and gave him his change. His hand lingered on hers a little longer than with usual transactions.

"Nadia," Isaac cleared his throat, "I realize these aren't the best of times in Benwood, but I was wondering if you might want to have dinner with me sometime?" He studied her carefully. "Or lunch?"

"Oh," she tucked a wisp of hair behind her ear. "I would… yes, I would like that."

"Tonight?"

"I'm sorry, Isaac. But tonight I've promised to work at the Red Cross tent down by the mine. There are still folks waiting to hear about their loved ones and rescue workers still need a cup of coffee and something to eat." She took in a deep breath. "You could join me," her voice lowered to a whisper, "if ya'd like to."

He smiled. "That's not exactly what I had in mind but I'll be there. What time?"

"Six o'clock."

He winked. "I'll see you there."

The Red Cross tent was a flurry of action when they first arrived. Nurses were dressing rescuer's cuts and scrapes, while other volunteers kept the coffee, tea and food prepared for whomever might need it—workers or civilians. Nadia was in charge of making bologna sandwiches, wrapping them in wax paper, and stacking them on the far edge of the table so they would be easily accessible for anyone in need. Suddenly she felt Isaac by her side. "What can I do to help?" he asked.

"You can slice the bread," Nadia replied.

He watched her for a second and began imitating her actions. He was standing near enough for the breeze to carry her warm citrus scent to him, and felt a strand of her silk-soft hair glide across his arm. She glanced up at him and as their gaze connected, something seemed to pass between them, something that shook her entire body and made her heart jump like a hooked catfish. Despite the noise and activity around them, the other volunteers and nurses chattering, the world shrank to just the two of them for a brief moment. Nadia blushed and dropped her eyes, hoping he didn't notice her cheeks turning bright red.

"Nadia," she heard a woman whimper. "Both of my boys are in there."

She turned to face the woman and realized it was Martha, her neighbor that lived at the mouth of Flanner Holler. "I'm so sorry, Martha." Nadia put her hand on the woman's thin shoulder. She stiffened beneath her touch. Her face crumpled and tears glittered in her eyes. "I know they're dead."

Nadia understood this much was true. Everyone who was in the mine was most likely dead by now. They hadn't found one survivor. "Have they identified them, Martha?"

The woman shook her head. "No. Not yet."

"Keep the faith." Nadia attempted to comfort her, knowing that mere words could not relieve the pain and apprehension the woman was feeling.

Martha gave her a feeble smile. "I don't know what I'm gonna do." She glanced toward the sky. "Why did God allow the mouth of the mountain to clamp shut?" She asked, a black slough of despair overcoming her.

"I don't know," Nadia answered, having asked herself the same question repeatedly during the last few days. "You'll make it through," she promised. "No matter what happens your friends and neighbors will help you."

Nadia realized that the wails of grief that had been constantly swirling about in the air had momentarily ceased and she figured most folks had used up all their tears. She could see widows soothing widows and childless mothers consoling childless mothers. As the woman slumped away, Isaac turned to Nadia. "I feel so sorry for her."

"So do I. I remember when my mother left us and then when Daddy died. I wanted to crawl into a little ball, fall asleep and never wake up."

"What kept you going?" Isaac asked.

Nadia's brow knitted as she placed her hand on his arm. "I had to take care of my little brother."

"You're raising your little brother alone?" Isaac arched a questioning brow.

"Yes, I am." Realizing her hand still rested on Isaac's arm as if it belonged there, Nadia snatched it away, turned away, and hurried over to fetch more bread to slice. They continued to work, side-by-side, until one of the nurses asked Isaac to help unload supplies from the truck. Nadia kept glancing his way, admiring his strong back, passionate eyes and determination to finish every task that was asked of him. The man must have a sixth sense for knowing when she glanced his way. He turned his head to her, and their gazes connected. Her heart gave a bump, but she didn't look away. Instead, she smiled.

Once the sun had fully tucked itself behind the mountain, the head nurse thanked Nadia and Isaac for their help and encouraged them to go home and get some sleep.

"Can I see you again, Nadia?" Isaac asked. "Maybe even take you to dinner and we can get to know one another a little better?"

"I'd like that," she answered. "But not until every man in this mine is accounted for."

He leaned forward and kissed her right on the tip of her startled nose. "I can wait," he whispered.

Then he saw the fire in her eyes as her fist struck his right cheek.

"Damn!" He rubbed his jaw. He knew he'd lost his ability to automatically filter his vocabulary, and Southern ladies demanded manners. "I'm sorry," was all he was able to sputter.

Charleston, West Virginia

1982

"The best way to meet a woman is in an emergency situation, like if you are in a shipwreck or find yourself right slap-bang in the middle of a flood or a tornado," Jack assured.

"A shipwreck?" I reverberated dubiously.

"Yep, it is very well known, Dee, that tragedy brings folks together." He nodded confidently. "Just look at us now. I'm seated at the kitchen table, sipping tea with two beautiful women at one o'clock in the morning and it's all because of an emergency situation."

"I agree," Nadia inserted. "If it hadn't been for the mining disaster, Isaac wouldn't have come to Benwood and I wouldn't have met him. It was a happy coincidence—happenstance, some might say."

I shrugged. "I don't know. Popping a man in his jaw doesn't seem like the best way to woo him, Nadia." I flipped my hand nonchalantly. "I know I'm not an expert on dating or anything, but didn't you overreact? Maybe just a little?"

She chuckled under her breath. "I definitely overreacted, but I had a bad experience in the past and when he bent down and kissed my nose, my mind involuntarily flashed back to our cabin up Flannery Holler."

"Why? What had happened?"

Nadia ruminated on this question for a few seconds, as

though trying to find the right words to convey to me. "A few weeks after we buried our daddy, William and I were working around the cabin on a Saturday evening. At the time, I was fourteen and William only ten. I had been avoiding everyone because I was afraid that folks would start saying that a fourteen-year-old girl shouldn't be raising her little brother all alone. I didn't want us to get split up and was concerned someone would try to 'find homes' for us."

Jack expounded, "You have to remember Dee, that we didn't have social services like we do in today's time."

"Exactly," Nadia agreed. "Anyway, a neighbor named Joe Scruggs suddenly popped in to visit us. He was drunker than a skunk and started saying things like, 'Nadia you're gonna need a man 'round to take care of you and the boy.' I kept tellin' him that we'd be fine but he kept goin' on and on. After a few minutes of arguing with him, he grabbed my head and forced his tongue into my mouth. He tasted like tobacco and moonshine and his body reeked of body odor. It was disgusting! I pulled away from him and told him to get out. But he was drunk and wasn't thinkin' clearly. Ultimately, he kept pushing himself on to me and started groping my body. I started fighting him off and William did too. He was a big fella, a little over six feet tall and heavy too. It took all we had to get him out of our house. We were hitting him with things and yellin' at the top of our lungs. Finally, he came to his senses... well, I don't know if he really came to his senses but he left. I understood how close I had come to being violated by him." She shivered at the notion.

"That's disgusting!" I murmured.

"Men like that should have their dilly winks chopped off," Jack proclaimed.

Dilly winks? I pondered, before turning my attention toward Nadia. "Did you stay there at the cabin?" I asked. "I wouldn't have hung around for it to happen again. I'd be too afraid."

"I was shook up, that's for sure," Nadia confessed. "After he left, I cried for a long time. I cried for Daddy and I cried for us. Then I realized the only thing I could do... the only place I knew

I could get help was at Mr. and Mrs. Bentzes' house. So, William and I packed up a few clothes and took off walking in that direction just before sundown."

"I am assuming they took you in?" I looked at her expectantly.

"They welcomed us with open arms." Nadia's lips trembled slightly at the memory. "They were so very kind to us. Really, Dee, they treated us like we were their own children and Mr. Bentz even got me the job working at the Cooey-Bentz Building selling dry goods. They were very generous."

I glanced at the clock hanging above the sink and pushed myself away from the chrome-and-vinyl dinette table. "Sorry to break up this party, but it's late and I need to go home, check on Gabby and get some sleep. Will you tell me another story when I stop by tomorrow… actually it's already tomorrow," I glanced at the clock again. *Two o'clock in the morning.* "I have classes today but I'll stop by this evening to see how you're doing."

Jack stood up too and snapped his fingers to wake up Max who was sleeping at his feet. "Yeah, we best mosey on home too. I'm glad Bing returned and I'm really glad I didn't have to get into a fight tonight."

Nadia and I laughed and nodded in agreement.

"C'mon by this evening and I'll tell ya'll another story about my early years."

"You bet ya!"

Burying the Dead

Benwood, West Virginia
May 5, 1924

Nadia could see, across the way from where she stood, Isaac respectfully clasping his hands behind his back as the services began. Twenty-two of the deceased coal miners were being buried side by side at Mount Calvary Cemetery. The funeral service was conducted in English first, followed by Polish and now it was being spoken in Italian.

She occasionally glanced in Isaac's direction but considering the fact that she had punched him on the jaw didn't make things easier. Nadia still cringed every time she thought about that moment. She had barely been able to make eye contact with him since that night and it took all her strength and courage not to turn around and run away from the services. When he had stopped in the store the day before she had darted to the ladies' room and directed her coworker, Wanda, to tend to him so she could avoid him and now, there he stood, just a few yards away from her.

Widows were weeping openly and sorrowfully. The variety of languages that she could hear in the prayers sounded dreamlike, almost as if the women were chanting. It didn't matter that Nadia didn't understand the languages, she knew what their words meant and she fully understood the sorrow and fear they felt deep in their hearts. Occasionally, she would recognize a name in their prayers—James, Salvatore, Wasyl, or Domenico

—husbands, fathers and children. Expressing grief, she decided, was universal. Once the last, "Amen" was uttered, she slowly walked along with the women, older folks and children as they moped mournfully across the expansive lawn.

"Nadia," she heard Isaac say as he closed in behind her. "May I speak to you for a moment?"

She flushed with embarrassment. "Sure. It's so nice to see you, Isaac." She smiled impishly.

He motioned to a bench resting beneath a large oak tree.

"Look Nadia, I need to apologize for what I did… I mean… when you smiled at me my brain suddenly felt like warm molasses… and it doesn't matter that you have a mean right hook. I can take the punch." He offered her a genuine smile.

"Isaac, I feel like I owe you an explanation of why I acted the way I did when you kissed…"

He interrupted her. "No, you don't. I owe you an apology. I don't know what came over me. It's just that you looked so beautiful, and I… I don't know…I just couldn't help myself." His dazzling, sage eyes drooped, causing him to look like a lost puppy dog. "Do you think we could start over?"

She shifted her weight on the bench and faced him. "Isaac, I really need to explain why I overreacted."

"You do not have to explain anything to me, Nadia. I should not have kissed you without your permission."

She sighed. "I'd still like to explain, Isaac. I am heading over to the Bentzes' house to help fix care packages for some of the families and then deliver 'em. Would ya like to join me?"

"Thank you. I would love to accompany you."

At the Bentzes' house, which was a charming white two-story bungalow located on Jacobs Street, Nadia and Isaac helped Mrs. Bentz pack up boxes stuffed full of homemade breads, a generous slab of ham, jellies, canned beans and a dozen eggs, the latter of which Nadia had brought from her house, into boxes Mr. Bentz had brought home from the store. It was enough to feed each of the families for a few days. It was a simple gesture but it would be greatly valued by those who were now attempting to pull their lives back together. Nadia knew Mr. and Mrs. Bentz had enough with a little left over to spare, but when it came to taking care of folks in the community they would often go without to help others.

The spring sunshine streamed through the kitchen window and spilled across the wooden plank floor and cheerful yellow walls offering Nadia a smidgen of hope. The last few days had been like living through a nightmare of pain, empathy and, at times, pure horror.

"If you'd deliver these to the DiGiorgios, the Dulas, and to Martha Malyska I'd greatly appreciate it," Mrs. Bentz told them as Isaac placed the care packages in the back of his truck. "And tell them we are keeping them in our thoughts and prayers."

"Why does Mrs. Bentz go through all this trouble?" asked Isaac.

"It's a simple rule really." Nadia attempted to explain, "If you and yours are safe then you comfort those who need consolation. Mr. and Mrs. Bentz believe that ya have to put your hands into your prayers. It's not just your prayers, it's your deeds that count, too."

"It must be a small town thing," Isaac replied. "I don't know who my neighbors are back in Charleston."

"You don't know your neighbors?" she asked, trying to wrap her head around the notion.

"I do not. I moved to Charleston two years ago and started working for the newspaper. I spend a lot of time going from town to town to cover events, and…" he shrugged his shoulders, "I just never met the people who live on my street, well other than my landlord, Mrs. Carrington. She appears to be a nice person."

"It must be lonely."

"Sometimes."

"Although living in a small town has a disadvantage, too."

"Like what?"

"Well, for example, when my mother left us we were the talk of the town I can assure you."

"How long ago was it that she left?"

Nadia expelled a long, shaky breath then scrunched her brow, trying to recall the days after Rebecca had left them. She had tried hard to forget about that time in her life, but now she wanted to share her sorrow and bring back the heart-crushing moments. "It's been ten years ago now. My mama packed up her sparse belongings and crammed them into a satchel. She put on her best dress and strolled straight out the door. I followed behind for a bit, so Mama wouldn't see me, as she promenaded a half-mile to the mouth of the holler. She didn't even look back and my eyes clouded with tears as she disappeared around the bend in the road. I was eight years old at the time and I remember sleeping on the porch after Daddy and I watched in silence as she left. I kept waiting for her to come home. I couldn't understand why she would leave me and I couldn't stop crying. My daddy brought a pillow and a blanket out to the rickety entryway and we lay there for hours, staring up at the stars and weeping together. I asked him, 'She ain't comin' back is she, Daddy?' Her daddy choked up. 'Nah, Sweet Pea. You're mama ain't comin' home,'" Nadia blinked back the tears that were teeming in her eyes and she gazed into Isaac's face. "My daddy was a good man."

Isaac simply nodded, not being pushy, not saying a word, but Nadia felt the sympathy emanating from him nonetheless.

"And I don't think of her as my mama anymore." She shook her head marginally. "She was the one who decided that."

Benwood was a quaint little town, despite the tragedy that just unfolded, with cobblestone side streets and well-kept storefronts. After calling on the families, Isaac parked his truck alongside Water Street just as the light was dwindling. "Would you care to take a stroll with me, Nadia?"

"I'd love to." Nadia felt her spirits lifting again. Having seen the faces of the grief stricken families when they had delivered the food had helped. Realizing how good it made her feel inside when she helped those in need had been the medicine she had been searching for ever since the morning the mine had exploded. Hope had been found in the midst of all the heartbreak that surrounded her.

The night was filled with the scent of gardenias and the shrill of cicadas hiding in the deep brush down by the river. "Thanks for tagging along with me today, Isaac."

"Thank you for asking me to join you," he replied.

Nadia felt his hand brush against hers, and then once more. He finally took her hand in his. She caught hold and they gently intertwined their fingers. It felt nice just to walk along in companionable silence with shadows of overhanging sycamore and willow leaves gliding overhead.

"Isn't it remarkable how good you feel once you've done something to help your neighbors?" Isaac seemed to suddenly realize. His words were like a sweet breeze flowing over her.

Nadia laughed a friendly laugh that seemed to pour over Isaac in melodious waves. "Yeah, it makes my heart sing."

They walked underneath the scattered lamplight, Nadia so close to him that he could smell the faint scent of citrus. He heard her sigh, and hoped it was a sigh of contentment not a sigh of despair. Finally, he got up the nerve to ask, "Nadia, if I kiss you, are you going to slug me in the jaw?" The words rolled off his tongue like ice cream on a warm summer day and made Nadia blush.

She giggled and turned to face him. "I reckon you could take the chance." She closed her eyes and felt his lips brush gently against hers and this time her heart soared unfettered with excitement and joy. *Southern men really are no match for any other,* she figured.

Charleston, West Virginia

1982

"Did you slug him?" I asked.

"No! We had quite a remarkable evening and the following day, he picked me up and drove me out to Greenwood where they held a burial for the miners who were burned beyond recognition. It was the saddest thing I've ever witnessed. They buried these men in a mass grave…" her voice trailed off. She shuddered. "It was disheartening."

"Why weren't you living at the Bentzes' all the time when the Benwood mining disaster occurred?"

"We lived with them for over a year and when they got word that Mrs. Bentz's parents were immigrating to the United States, I decided I didn't want to be in their way anymore. I knew they'd need the space for 'em and figured William and I would just be in their way. Dee, there were a lot of immigrants coming to America during this time. Mr. and Mrs. Bentz had come from Germany a decade earlier, and many of the folks who lived in Benwood had just immigrated from all over the world." She paused momentarily. "We really had quite a diverse community."

"The American dream," I murmured.

Nadia smiled. "It sure didn't seem like it to the folks who buried their kin after the accident."

"I'm sure."

She flipped through her memory album and carefully chose

a picture. "This little boy was an immigrant from Poland. He was a loader. If you recall, I went with Mr. Bentz to tell his mama that he had died in the accident."

I stared at the child for a long time. Even though the black and white photograph wasn't very clear, I could tell he had a black eye, slicked back hair with a straight part on the side and he was wearing a suit. I estimated his age to be around nine or ten years old.

"You should have seen him in color. You can't tell it here but his tie was bright yellow and his eyes were as blue as the sky on a summer day. His name was Jan Piehowicz, and his poor mama was never the same after her little boy died."

"I can't imagine."

"It was a sad day," Nadia guaranteed. She carefully positioned the photo in her album and glanced up at me. "Anyway, enough about my life. How was your day?"

"My classes went well and I stopped and had the tire pressure checked at the gas station." I shrugged. "Nothing exciting."

"Are ya hungry? I fixed a poor man's supper for this evening."

"A poor man's supper?"

"Yep, I have a pot of pinto beans with ham hocks and they've been simmering on the stove all day, and there's a mess of cornbread ready to put in the oven."

"Yes! That sounds wonderful!"

"Jack should be stopping over soon, so we'll wait for him if that's okay with you."

"Absolutely," I replied. "And Nadia, I'm really happy that Bing came home. You must be relieved."

She tittered. "I am thrilled beyond belief."

Just then we heard a knock at the door. "C'mon in. It's open," Nadia called out.

Jack and Max made their way through the living room and into the kitchen where he handed a bouquet of daisies to Nadia.

"Oh, ya didn't need to bring me flowers, Jack."

"My mama would roll over in her grave if she thought I was showing up to a meal empty-handed."

Nadia accepted them with a smile and placed them in a Mason jar before filling it half-full of water.

"What smells so good?" Jack asked.

"Pinto beans and cornbread," I informed him.

"My favorite," he replied, settling down on a kitchen chair. "So, did I miss any of your stories about Benwood?"

"Nothing exciting," Nadia assured.

"Isaac kissed her," I butted in.

"Did she slug him? She has an ornery streak, you know."

"Not that time," I relayed.

"Do you remember what you felt when he kissed you?" I asked. "Did it make your knees buckle or your heart flutter?"

She grinned. "I remember how I felt, but not what I felt."

Jack grunted. "Isn't that a quote by William Wordsworth?"

"I don't know," Nadia thoughtfully revealed.

"It's darn near close. I think he once wrote, 'Remembering how I felt, but what I felt remembering not.'"

"Jack," I pointed at him, "how is that you are such a literary giant?"

He looked at me with all the seriousness he could muster. "I read books."

I offered a sarcastic smile.

Nadia slid a cast iron skillet into the oven and proceeded to cut slices of ramps into tiny pieces. She placed mismatched bowls and plates on the table in front of us along with two sticks of butter. "Milk or sweet tea?"

"Milk," Jack and I answered in unison.

Just then I recalled what I needed to tell Jack. "I stopped and got my tires checked at the gas station today. Thanks for letting me know the one was losing air."

"You're welcome, Dee, but I would have checked it for you."

"You don't have to take care of my car for me, Jack."

"I know I don't. But I'm not doing much of anything else, so I don't mind," he promised.

I noticed Nadia fill up a bowl of water and set it on the floor for Max, before directing her attention to Jack. "Did ya get into anything today, Jack?"

"No, Max and I went home and caught a couple winks, I went down to the pawn shop and sorted through a few more boxes then we came home and watched *As the World Turns*."

"You watch soap operas? Really?" I teased.

"I do." He squared his shoulders. "Don't you?"

"Sometimes," I admitted.

By the time Nadia sliced the cornbread, filled our bowls full of juicy beans and poured us each a glass of milk, I was starved. She said grace and we dug in like we hadn't eaten in days. A full ten minutes passed before anyone said a word. "You are a fine cook, Nadia. Do you mind if I have another helping?"

"It does a cook's heart good to see someone going back for seconds," she replied.

He peeked at her. "How about thirds?" Jack asked, treating the meal as though she had gone gourmet on him.

Nadia and I laughed out loud.

Just then I noticed a fly land on top of a slab of butter on Jack's plate. He shooed it away with his hand.

Nadia blushed. "I'm so sorry, Jack. I have a hole in the screen." She pointed to the kitchen window.

"Oh," he clucked his tongue, "that little bugger won't eat much, and if you remind me later I'll fix the screen for you."

I could tell that Nadia was embarrassed about having a fly buzzing around the kitchen and landing on her guests food, but I also felt a bit of admiration for Jack. He didn't miss a beat.

"How about you tell us what happened after Isaac gave ya a smooch," Jack suggested, sopping up the last of the bean juice with his cornbread. He pushed his plate away and laced his hands across his stomach, waggling his brows conspiratorially.

"Oh," her hand rose to cover her mouth and she giggled.

"I like that idea." I picked up a short stack of dishes. "I'll clean up and you can fill us in on the intimate details."

"I'm skipping on to the next part of the story. There's nothing more that I want to share about the kiss." She blushed like a schoolgirl, before pointing to me. "Don't use my new set of flour sack tea towels. They're too good to use up."

"Okay." I turned and winked at Jack.

Nadia started clearing the table, even though I protested, and we all juggled for space in the small kitchen. I noticed Jack reaching for aluminum foil, while Nadia was searching in a drawer for a clean cloth. They turned at the same time and collided. Her eyes locked on his broad chest before she lifted her chin and gave him an awkward smile. Jack's eyes lingered on her for a long moment, so I cleared my throat vociferously to remind them I was standing in the room.

"You both go sit down. There's not enough room for all of us by the sink and I'm cleaning up. Go!" I made a swooshing motion with my hands. "Tell us another story, Nadia."

Despair and Hope

Benwood, West Virginia
May 25, 1924

"Nadia," Mr. Bentz called out. "Could you come to my office for a moment?"

"Yes, sir." She finished folding the cotton towel in her hands and placed it on top of the stack on the table before hurrying to his office.

"Is everything all right?" she asked.

"Please, have a seat." He motioned to one of the wooden chairs on the opposite side of his desk.

She dropped down into it and stared at him with wide, inquisitive eyes. She could easily ascertain by the grim expression on his face that something was definitely amiss. *Please God. Don't let there be something wrong with William.*

"Nadia," Mr. Bentz began in an excessively consoling tone of voice, "I have some bad news to share with you."

"Has something happened to William?"

"No." His lips formed a taut straight line. "I've just come back from the Bank of Benwood where I spent the afternoon talking to Mr. Roberts. Apparently," he cleared his throat, "someone has embezzled all the money from the relief fund for the widows here in Benwood."

Discouragement flooded through her. "Do ya mean the money folks have been sending from all over the country?"

"Yes, and it was quite a substantial sum."

Her mouth gaped open. "That's wicked!"

He lowered his eyes. "Unfortunately, they are ninety-nine-percent sure that James is involved."

Nadia let this sink in for a moment. She faltered. "James? Do you mean Mother's husband?" Her cheeks flushed with disgrace and she had a hard time looking at the sympathy—or perhaps it was pity—in Mr. Bentz's eyes. She didn't want people to feel sorry for her. Not again.

"I'm sorry, Nadia. I know you don't have strong feelings for Rebecca, and quite frankly neither do I, but the news will be breaking soon and I wanted you to know before you heard it on the street."

Intruding scenes from years ago flashed relentlessly before her eyes. *"That's the little girl whose mother ran off with another man and left her husband and children behind."* They'd say as she walked past. *"Poor girl, first her mama was a tramp and now her papa just ups and dies on her."* Whisper layered upon whisper built to a crescendo. *"I have no idea how she thinks she'll be able to raise her little brother on her own."* Nadia took in a deep breath and closed her eyes. Cold dread filled her heart.

"Let me give you some unsolicited advice," Mr. Bentz pinned her with a listen-to-me look. "Even though you don't respect Rebecca after what she did to you and your family, she is still your mother. Therefore, if anyone in town says something disrespectful about her I would recommend that you simply tell them flat-out that *you* do not gossip and spread rumors about anyone and that you'd kindly appreciate it if they would show you the same respect. Everyone adores you, Nadia. Her exploits, nor her husband's deeds, are a reflection of the wonderful young lady you have blossomed into. You are not responsible for any of this, so don't think for one moment that you need to apologize to anyone."

Nadia's lip trembled. "It's just so embarrassing." The bitterness of years boiled to the surface. "I feel ashamed."

"That is exactly what I am talking about. You are not responsible for anyone's indiscretions but your own. Please, just hold your head up high and remember that no one can make you feel inferior without your permission."

She buried her head in her hands. "Oh, Lordy," she murmured. "Please send help."

Mr. Bentz handed her a freshly ironed handkerchief. After several long moments he asked, "Would you like to fetch William and stay at our house tonight?"

Nadia shook her head.

"Can I drive you home?"

"No," she whimpered, "I rode my bike to work. William and I will be fine back at the cabin." She honked her nose and fumbled with the soiled hanky, "I'll wash this up and return it to ya."

His lips turned up in a sad smile. "You keep it, dear." He drummed his fingers on his desk. "You know that Mrs. Bentz and I love you and William like you were our own."

"I wish we were yours," Nadia sniffled.

He rose from his desk and walked around to give her a crushing bear hug. "Don't forget what I told you. Hold your head high, Nadia. You have no reason to feel humiliation because you are not responsible for anyone else's conduct."

"Even when they're kin?"

Mr. Bentz offered a musing smile. "Particularly when they *are* kin."

Nadia carefully mounted her bike, making sure she didn't show off her undies to the entire world, hoisted her leg over the saddle and lined up her feet for the pedals. Her feet started pressing firmly on the pedals with great force as she determinately headed home. The very thought of another scandal involving her mother made her so angry that she didn't realize she was pushing hard and recklessly. Her heart was beating wildly and she could feel it pounding in her chest. Perspiration dripped down her back as disturbing memories flashed through her mind:

> *"Where's mama?" she asked her daddy.*
> *"She won't be comin' home," he put in plain words. He choked back the tears.*
> *"Why?" Nadia's bright eyes studied him expediently.*
> *"She's found a better man."*
> *"A better man?" Nadia echoed.*
> *"Yeah. It'll just be me, you, and William from now on."*

Nadia had tried to understand this, but couldn't quite wrap her mind around it. *"Daddy, there's not a better man than you."*

She recalled her daddy crying for nights on end, and the way his face turned the color of a strawberry in summer, when he explained to folks why his wife, Rebecca, wasn't around anymore.

Nadia was ashamed. *"I hate her!"*

Suddenly, the bike's front wheel caught suddenly in a pothole, and she went hurtling to the ground with a booming crash. She cried out in pain as her knee scraped on the gravel as she slammed against the road. Nadia glanced down only to see the tail of her dress hitched up above her waist.

"Nadia! Are you okay?" She turned to look and saw Isaac slamming the door of his truck and she could hear the concern in his voice, then a moment later he was lifting the bicycle off her with one hand. Her knee was throbbing but what was worse was the disgrace she felt at having Isaac witnessing the entire event. Her looking like a total sweaty fool with a scraped knee and her bloomers fully exposed to any prying eyes.

"I'm fine!" she insisted, sniffling with pained tears. She smoothed her skirt down to cover her bottom.

He bent to help her up. "I'm driving you home," he said, as he tossed her bike into the bed of his Ford truck. He opened the door and she clambered up into the passenger seat. Isaac rushed around the truck and slid into the driver's seat. He looked over at her. "I was on my way to the store when I saw you speeding like a demon down the road. Are you upset about something?"

"No," she stubbornly crossed her arms in front of her chest. "Just take me back to the store."

"Weren't you going home? It looked like you were headed that way."

"Isaac," she barked, "I don't want you to see the dingy shack I live in." Exasperation made her curt. She couldn't believe she had vocalized the words—actually said them out loud! *What is wrong with me?* Still, she didn't want him to see the scraggly lawns with weeds cropping up everywhere, nor the bedraggled, bumpy holler where children played in patches of dirt because the grass had given up all together. She was ashamed to show him the grubby place she called home.

"Nadia," Isaac said with a comforting declaration, "I don't care what kind of house you live in." He glanced over at her. "I love you."

"You love me?" she spoke softly, wanting to make sure she heard him right.

He nodded and smiled, shoving a spray of lavender across the seat of his truck. "I sure enough do."

She felt her heart warm up like an electric light. The feeling of anger and resentment that she felt toward her mother disappeared in an instant and was replaced by soothing warmth —butterflies fluttered in her stomach. The way she felt about Isaac frightened her, she wasn't sure if she was ready to put herself out there—vulnerable, with no defenses. Nadia had worked hard to build a wall around herself after her mother left them and vowed to never trust anyone again—then, her father dying and leaving them alone too.

She glanced over at him and took in his tasseled hair, broad shoulders, dimpled chin and mesmerizing green eyes. She plucked up the bouquet of flowers and breathed in the sweetness. "I love you too," she whispered faintly.

Isaac's heart skipped a beat. "Can I ask you something, Nadia?"

"Sure."

"Will you marry me?" His voice was deep, velvety and full of soul.

Nadia took in a deep breath. A glimmer of hope sparked inside her. "Marry you?" She could feel the attraction hanging over them—thick like a humid West Virginia summer night.

"Yes, ma'am." He winked. "I made a wish and you came true."

Nadia couldn't believe his words. She wanted to, but it seemed like the stuff that dreams were made of, and she didn't want to end up disappointed. She took in a lungful of cleansing air. "You do realize I must take care of my brother, William, right?"

"I understand completely. I intend to ask him for permission to marry his sister since he would be living with us."

Her breath caught and her heart tripped. Not only did he want to marry her, he was also willing to help with William. Her mind whirled. "Well," she breathed out slowly, "you may not be so keen on marrying me once you read the newspaper tomorrow."

His brow arched. "Oh? Why so?"

"Mr. Bentz just informed me that my mother's husband, James, embezzled all the money from the widow's trust fund." She chewed on her lower lip.

"I know." Isaac shrugged. "James has no more chance of getting out of this than a grasshopper in a chicken house."

"You knew?" Nadia felt like throwing herself out of the moving truck just to escape the humiliation.

"I'm a reporter, Nadia. But don't you worry. I won't be sending anything to print concerning your mother or her husband. Anyway, what does that have to do with you?"

She stared out the side window, not wanting to meet his gaze. "Nothing," she replied in a low voice. "She's just disgraced our family. Again."

He drove his truck to the side of the road and shut down the engine. "Nadia, please look at me."

She turned and lifted her eyes to meet his.

"Your Mama will be paying the high cost of low living, but

it doesn't mean that you have to. Sin is expensive and she'll pay the price."

"I reckon."

"You reckon you'll marry me or you reckon that you don't need to worry about what folks say about your mama?" Isaac teased.

Charleston, West Virginia

1982

"So, did you say yes?" I asked, placing the clean bowls back into the cupboard.

Nadia squinted her eyes. "Of course I said yes, Dee."

"Right," I murmured. "I knew that. Did ya'll stay in love?"

A sassy smile wrapped up around her cheeks. "Indeed, we did. Life wasn't always easy but we made it through."

Jack belched. "Excuse me. I may have eaten too much." He patted his belly before shifting his weight, causing the chair to groan in response. "Just listening to you talk about how great your husband was puts me in mind of my third wife, Marsha. We were married a couple of months before the newness started to wear off, and we were starting to get under each other's skin. One day," he chuckled to himself, "we were driving to see her parents, and we were arguing in the car. As we drove past a barnyard full of pigs and donkeys, I couldn't resist saying, 'Relatives of yours?' She smiled right back at me and said, 'Yeah,' without skipping a beat, 'my in-laws.'"

Nadia and I started laughing out loud.

"Serves you right," I quipped.

"I know." He shook his head. "I wasn't the easiest man to get along with back then and she was a feisty woman. Marsha was the third horse in the trifecta. She was like the Kentucky Derby of women I've disappointed in my life." He dipped his chin apologetically. "I'm sorry, Nadia. I didn't mean to interrupt you."

"Wait," I interjected. "How many times have you been married, Jack?"

"Only three. After Marsha left me I figured I'd best go it alone for a bit."

"How long ago was that?" I probed.

"A long time ago," he guaranteed. "I've grown up a lot since then."

I murmured, "I would hope so."

Jack grunted but ceased to comment. Abruptly changing the topic, he addressed Nadia. "You know, I've been thinking. Why don't you fix-up some of the old furniture I have at the shop to make them more appealing? I could haul one of the old tables over here to your house, you could paint some fancy designs on it, like you have here in your house, and then I could sell it. I figure I could display the pieces in the bay window at the shop. Plus, you could make some extra money too. What do you think?"

Nadia seemed to consider his proposition carefully. "Maybe. I'm not a professional though."

"Oh, that don't matter. You have a knack for it."

Nadia's cheeks flushed. "I'll roll it over."

"Good," Jack nodded with approval. "Because, I've been thinkin' about hiring a married man to help me out at the shop."

"Why would you want to hire a married man?" I asked. "Why not a woman? Do you think a woman is not smart enough to run a cash register? I mean, really, Jack."

"Oh, not at all. It's because married men are used to obeyin' orders, gettin' shoved around, keepin' their mouths shut, and not poutin' when they get yelled at." He smacked his leg pretentiously before laughing out loud.

We joined along in his contagious fit of hilarity.

"You should hire Nadia. She has experience working in a store. Remember? She worked for Mr. Bentz," I suggested to Jack.

"I worked at the Diamond Department Store too," Nadia informed us. "Back when I started working there it was called The

Diamond Shoe and Garment Company. The name was changed in 1927 when the store moved locations."

"Really? I love that store," I enthusiastically responded.

Jack exclaimed, "It's a fancy place." His attention turned to Nadia. "Would you consider working at the pawn shop?"

"And take orders from you? I don't think so," she waved the idea away.

"Sounds like a good idea to me. You could add a little class to the store, Nadia."

Jack's jaw dropped. "What's that supposed to mean, Dee? You don't think the pawn shop is classy?"

I responded with a very unladylike snort. "It could use a woman's touch, Jack. That's all I meant."

Nadia shyly interrupted, "I have been working on sprucing up a table I found at a yard sale."

"Where is it? I didn't know you were working on something new." My eyes darted around the kitchen and through the open doorway leading to her living room.

"Come out to the porch and I'll show you." She motioned with her hand indicating we should follow her. Jack and I followed her and Max through the house and out to the front porch. Jack and I heard Nadia's distraught gasp, followed by a bird screeching and a dog whimpering, before we saw what was causing the ruckus. Apparently, as Nadia and Max slipped out the door and onto the porch, Bing decided to land on top of Max's head, causing the old dog to jump, which triggered a can of blue paint to topple down from the table Nadia had been re-finishing. Max was now enveloped, from head to toe, in a most striking shade of cobalt blue.

Jack and I stared at Max, who was now shaking the wet paint from his fur, causing tiny splatters to cover the porch floor, furniture and our legs.

"Oh, Lordy," Nadia cried out. "Let me fetch a towel."

Jack tried, as best he could, to wipe the tinted paint from Max's stout body. He handed Nadia the stained towel and asked, "Is this a water-based paint? I hope so."

"It is."

"Good." He grunted. "I guess I'll take the old fella home and give him a proper bath." He motioned to the splashes of paint speckling the area. "Sorry."

"'Taint your fault," Nadia assured.

"Ya'll stop on by the store tomorrow if you have time," he murmured.

"Will do."

We watched Jack and Max hurry down the street before Nadia found some old cloths and we started scrubbing the floor and other areas, hoping to find all the spots before they had a chance to dry. Once we were satisfied with the cleanup job, Nadia turned to me and suggested, "Let me tell ya about the day I visited Rebecca."

"Is that the next part of the story?"

"Yep."

Leaving Benwood

Benwood, West Virginia
June 28, 1924

Nadia slipped on her one and only store-bought outfit. It was a robin's egg blue tunic dress that fell straight from the shoulders over a low flattened bust. A line of lace adorned the neckline and hem, and she had even purchased a strand of fake pearls to complement it. Sure, Mr. Bentz had insisted she paid only the invoice price—after she had turned him down on this fine offer to buy it for her—but this dress was hers. She picked it out. She had been making payments on it for weeks, even when Mr. Bentz insisted she didn't need to. She wanted something nice—something that she had bought with her own money. The dress made her feel independent, like she didn't need to rely on anyone else. It was a silly thing for a dress to do, but it did.

Nadia wanted Rebecca to know, to realize, that she had done just fine for herself even though *her mother* had abandoned her when she was ten years old. *Yeah, we did just fine without ya.* She squared her shoulders and knocked on the front door of the two-story Victorian situated at the edge of town.

When the door swung open, Nadia stared at the woman on the other side of the threshold. Her mother, Rebecca, had dark circles under her eyes. Her hair was uncombed and it was easy to determine that she had not been expecting company.

"That's a pretty fancy dress, Nadia. I guess that reporter fella

I heard about, is carrying you around on a silk pillow, eh?" Rebecca said, motioning for Nadia to follow her.

Nadia's eyes flashed, full of anger. "Rebecca, I have a job. I've had one ever since Daddy died. How do ya think William and I have been able to afford to eat?"

"A job?" Rebecca repeated. "Working as a clerk? You call that a job? Nadia, believe me, you need to lasso the first fella who has a couple dollars in his pocket, and get out of this town."

Nadia stared reproachfully into the other woman's eyes. "We've been doin' just fine without you. Then you traipse into town and muddle everything up."

"What *exactly* do you mean, Nadia?"

Turning to her, white-lipped with anger, she furiously said, "Mother, haven't you shamed us enough? First you leave your family behind and now your husband, James, has been accused of stealing money from the widow's relief fund. His actions are repulsive."

Unaffected, Rebecca motioned toward the sofa. "Would you like to have a seat?"

Nadia sat on the edge of the sofa and tucked the tail of her dress around her hips.

"Nadia, you are young and don't understand anything about life. Not one person goes through their life without experiencing pain. Life's not easy and it's not fair." Rebecca rushed through her spiel. "I left because I wanted more out of life than living up a muddy holler and never owning a decent dress or pair of shoes. I knew if I stayed I would end up withering away and resenting your daddy." She paused to allow Nadia to take it in. "Just in case you aren't aware, your father and I had supper before we said grace."

Nadia felt a sudden jab of pain followed by an intense pounding in her chest. *Had supper before they said grace?* "Rebecca, are you saying that you were pregnant with me before you married Daddy?"

"Yes, that is the *only* reason I married him," she replied quite pointedly.

Nadia took a long moment to process this information. "Did you have supper with other men before *saying grace*?" Nadia asked with a note of sarcasm in her voice.

Rebecca stared at Nadia, rolled her eyes and replied through gritted teeth, "What do you think I am? A floozy?"

Nadia shrugged her shoulders. She grasped the dash of annoyance crossing the other woman's face.

Unexpectedly Rebecca retorted, "So what if I did?"

Nadia couldn't believe the words that had just escaped Rebecca's mouth. "Well, I'm askin' if my daddy was in fact my daddy, or if it could have been any number of men who live in Benwood."

"Oh, please." Rebecca dismissed her assumption. "It hardly matters now."

It hardly matters now? Nadia moaned. Her mind raced. She glanced around the room, noticing that Rebecca was either unpacking her things—or perhaps packing up her belongings. Boxes were flung about filled with clothes, shoes, and kitchen utensils. As much as her mind told her to get up and move, her body ignored the command. She sat there speechless for a long moment then finally it occurred to her. "You've been in town for almost two months now, are you just now gettin' around to unpacking or are ya leaving town again? It kinda looks as though you're in a hurry." Rebecca acted like she was up to no good and she seemed to notice that Nadia suspected something. Nadia saw a world of weariness and a lifetime of failure in the other woman's eyes.

"Yes," Rebecca replied brusquely. "I'm leaving town."

"With the relief money that James stole from the widows and children of Benwood?" Nadia asked, her voice plaintive.

"You have no idea what you are talking about," she answered flatly. "So Nadia, what is the reason for your visit today?" Rebecca tilted her head awaiting a response.

"I just wanted to let you know that Isaac and I are to be married and we'll be moving to his home in Charleston. William, of course, will be living with us."

"Well," Rebecca smirked, "I guess you did find a way to get out of this little town."

"Rebecca," Nadia pleaded, "if you know where James stashed the money I am begging ya'll to return it. Those women need that money to get back on their feet. To feed their fatherless children."

Rebecca's brow arched. "I have no idea what you are talking about."

"I don't believe you."

"I don't care," Rebecca snorted.

"Fine." Nadia fixed her gaze on the floor.

"Look Nadia, I know you don't love me like a mother and I understand why you feel that way. However, I have something I want to give you anyway." Rebecca rose and walked across the room, burrowed through a large crate and pulled out a diminutive box. "These are my memories of your daddy and of you and William. These are the photographs I took with me when I left. It's not much, but they're all I have left of the days when we were together." She handed the box to her daughter.

Nadia accepted the container but didn't open the box to examine the pictures. "Look, if you ever get desperate and don't have a place to live, you can move in with us. We'll be livin' in Charleston."

They stared at each other for a long drawn out moment. "Does this mean you have forgiven me?"

Nadia took in a deep breath. "I've forgiven you, but I'll never forget what you've done."

"Nadia, I know I've disappointed you as a mother, but please remember me."

"I will," Nadia vowed. And with that, Nadia stood, turned and walked toward the door—promising herself that she would never turn out like the woman she used to call her mother. She thought she heard Rebecca say, "I love you," before the door slapped shut behind her—or perhaps she just romanticized that she heard the words. Either way, it didn't matter anymore.

Charleston, West Virginia

1982

Nadia slid into the passenger's seat of my Camaro. Once she was situated, I revved the engine. "I still can't believe you offered to let Rebecca move in with you after all she and James had done. I thought about it all night long. I was even rolling it over in my head when I fed Gabby this morning. I mean it was *really* a generous gesture on your part."

"Isaac suggested it. He said kin was kin and you always take care of one another, regardless of what they've done."

I wasn't completely sure I agreed with him, but what did I know? My parents had always been kind to my siblings and me and had never caused us any embarrassment.

"Dee, I will tell ya one thing that's for sure. The box of photographs Rebecca gave me was one of the best gifts I've ever received. It held the only known picture of my daddy. He was my hero. I still treasure the old black and white photo to this day."

"I'm sure you do." I reflected for a long moment on the importance of saving old photographs. "Did you ever see Rebecca again?"

"No. That was the last time I saw her and I have no idea what happened to her."

"Did you ever wonder how her life turned out?"

"Of course," she replied flatly, seemingly not wanting to continue the conversation.

"Hey Nadia, I've been thinking about something you said

earlier this week. You mentioned that Isaac always looked at you like you were the only woman in the room."

"Yeah, he sure did. We had a wonderful life together. All the way up until he developed Alzheimer's disease… then he didn't know me."

"But you knew him."

"Exactly."

"Was it horrible, Nadia? I've heard Alzheimer's disease is really tough on the spouse and the children."

"Dee, it was bad. It started out with little things. For example, Isaac couldn't connect his thoughts. The words he seemed to be trying to express and the words that actually came out were entirely different. For example, one morning he pointed to an orange and asked me to hand him an egg."

"What did you do?"

"I handed him the orange and didn't say anything. Eventually, he stopped eating. In fact, it was as though he forgot how to chew his food. One morning I had cooked his favorite breakfast of scrambled eggs, bacon and toast. He had practically stumbled to the kitchen table and was still wearing his boxer shorts. I didn't care, mind you, because a man can eat breakfast in his boxers if he wants to, right?" She peeked at me. "It was just that he always got dressed the moment his feet hit the floor. Anyway, I poured him a cup of coffee and watched his trembling hand lift it to his lips. He gazed at me with uncertainty. His brow knitted and he looked confused. 'You are beautiful. What's your name?' For some reason my hand started shaking too. I took in a deep breath, unsteadily rested my cup into the saucer so I wouldn't slosh coffee everywhere, and looked straight into his eyes. I realized he didn't know me. So I said, 'I'm Nadia. It's a pleasure to meet you.'"

I tried to picture this in my mind's eye. "It must have been very hurtful."

"It was," she assured. "At first, the shock of watching Isaac's deterioration was horrifying. He only became totally lucid and free from the Alzheimer's that gripped him every now and again, so I asked his doctor a lot of questions and even borrowed some books from the library so I'd know what would be coming. Toward the end, Isaac lost a great deal of weight quickly and his skin seemed to hang from his face. I even had to buy him a pair of new shoes because his feet shrank by two sizes. Everyone insisted I have him put in a nursing home but I refused. I knew I could take care of him even if he didn't know who I was." She shrugged her shoulders. "I promised to love him 'until death do us part' and I did."

"You're very brave, Nadia. I'm not sure I could take care of someone who didn't recognize me," I confessed.

"Oh, you would, Dee." Her thoughts seemed to drift. "There was another day when he alleged I had broken into the house and was stealing his wife's things. He kept saying, 'Get out of here! Why are you stealing Nadia's shoes? I'm calling the police!' Foolishly, I stayed and he began throwing things at me."

"Had he ever been violent with you before?" I asked.

"Never. He had never even raised his voice to me during our entire marriage."

"So, what did you do?"

"I ran into the bedroom, shut the door, and pulled our wedding photograph off the wall. I stayed there for several long moments before returning to the living room. I gently explained to him that I *was* his wife. I showed him the photograph and told him over and over again that I was *his* Nadia." She sighed. "Then something snapped and he suddenly recognized me. He said, 'Did that woman leave? She was trying to steal your shoes.'"

Nadia fumbled through her handbag, pulled out her wallet and held a picture up for me to see. I glanced at it before quickly returning my focus to the road.

"This is one of my favorite photos, which is why I keep it close. This photograph is of Isaac and me in the summer sun, we're all dressed up because it was the day we said our vows." She traced a finger over the edges of her memory. "Mr. and Mrs. Bentz bought this wedding gown for me."

"You are beautiful and Isaac was a stud muffin." I purred like a kitten to emphasize my point.

Nadia laughed out loud. "Yeah. You can't tell it here but it was hot that July, and the rose I'm holding was ruby red. Just look at my smile, I was so contented and truly valued the moment." She stared at it for a little longer before slipping it deftly into her wallet. Her voice lowered to a whisper. "You should've seen us in color."

Even though it sounded like such a lame cliché, I said it anyway. "A picture is worth a thousand words."

"Yeah. I look at his photograph all the time, the memories come back to life and I don't mind." Nadia looked over at me. "Let me tell you something I've never divulged to a single soul, because I feel ashamed of how I felt."

My eyes darted toward her for a second. "Nadia, everyone feels ashamed of their actions or thoughts from time to time. It's perfectly natural and I promise I'll never repeat anything you tell me, unless you want me to."

She nodded. "Well, here's my confession." She gulped back the lump in her throat. "When Isaac died I was happy. I know it sounds terrible to say that, but I was glad that he would never have to suffer again, and I knew he rarely had any idea of what was happening to him. For that, I was grateful. You have to understand—it wasn't for me that I felt that way. It was for him. He is in heaven now and his body and mind are whole again."

I really didn't know how to respond because I had never watched a loved one slowly deteriorate physically or mentally. Finally, I replied, "I'm sure it was a difficult time in your life."

Nadia heaved a wearied sigh. "It was. However, he'll always be in my heart. Because when someone dies they live on in the people who knew and loved them. Mind you, I didn't grieve for him. I grieved for myself. I knew my life would never be the same without him, and I'd miss him for the rest of my life… every moment of every day."

"Did you have anyone coming in to help you with him? Like assisted care?"

"Yes, we had a nurse stopping by every day for the last

month or so of his life."

"Did you know that he was going to die? I mean... did he display any behaviors that made you think his time here on earth was coming to an end?"

"Yes, indeed. I could see the signs a-comin'."

"What did you observe?" I asked curiously.

Nadia carefully considered my question before answering. "Well, a few weeks before he passed his appetite decreased and his need for sleep increased. I had to start helping him more with simple things like getting dressed and eating. Then a few days before he passed on over, his level of consciousness declined, along with his blood pressure and then he started murmuring... and reaching in the air like he saw something or someone I couldn't see." She swallowed deeply. "Then the night he died, he couldn't swallow and appeared to be in a comatose-like state. I noticed his feet and hands were very cold when I crawled into bed with him and spooned up against his back. That's when he said his last words to me."

"What did he say?"

"He said, 'Remember me.'" She shrugged her shoulders. "As though I'd ever forget him." Her voice softened as the memory of that night returned. "That was the last time I heard his voice." A heartbroken sigh escaped from her lips. "I woke up around four in the morning and heard him take his last breath." Nadia's voice grew raspy—like mine always does when I'm trying to talk and fight back tears at the same time. "Then he passed on over to the sweet by and by. His soul and spirit left, but I held his body for a long time. I was unwilling to let him go. Now I realize how fragile life is and how important it is to live in the moment because we never know when that moment will be taken away from us."

"I'm sorry."

With a sad smile she added, "I was an old woman having to let go of the love of my life—my Prince Charming. Then once I thought it through I was happy for him."

I popped the gearshift into 'park.' "Well," I motioned toward

Jack's Pawn Shop, "here we are."

Nadia applied a coat of lipstick before admitting, "I hope I don't run into anyone from church."

"Why?" I didn't understand her reasoning.

"They might think I'm here to pawn something off."

"Nadia," I tilted my head, "if someone from church is in there, that means they are in the same place as you are, so you shouldn't be embarrassed if you see anybody. Plus, lots of folks stop in to peruse his shop. He has some great treasures."

"I guess you're right," she decided. "What do you think about Jack's suggestion that I decorate some of the old furniture he has here?"

"I think it's a great idea. You are very talented and you could earn a little extra money." I pointed at the deep bay window. "Furthermore, his shop could use some sprucing up. It would make it look sophisticated."

Nadia stared at the building. "The storefront is kinda plain, isn't it?"

"Oh yeah, it's very manly looking."

She giggled. "I'll think about it."

Jack's shop was located in a two-story building on Capitol Street. The large building was painted dark gray and had a deep bay window that he should have been using to display some of his finest antiques. Instead it was empty and bleak. Although, the thick window trim and French door for the entry were painted a bright white and the contrast was perfectly pleasing.

When we stepped inside Max came up to greet us. I figured Jack didn't need a bell to chime announcing a customer's arrival, because Max was there to do it for him. I gave him a lively rub on this head. "How are you doing, buddy?"

He jiggled his head, which I took to mean that he was doing fine, and slimed my palm. *Ugh.* I swiped my saliva-covered hand across the leg of my Lee jeans and attempted to talk myself into an accepting state of mind. *Dog slobber is not uncommon.* I chastised myself. *You pluck an apple straight from the tree; swipe it on the leg of your pants before taking a bite, right? How could that be sanitary?* I reminded myself. *It was only last week that you drank straight from the outdoor water hose.* Yes, I did. I wasn't going to dwell on germs at this moment, so I changed my train of thought.

The interior of the shop was dark and dated with century old cabinets, bookshelves, and curios lining the perimeters. Boxing memorabilia was displayed throughout the store on the faded walls. Floor space was limited to just enough passage for two people to walk side by side, and dingy brown carpet covered the entire floor. Boxes of old things were stacked against the wall beside the cash register and I noticed a few old porcelain vases in the two-dollar bin that hadn't been there when I stopped by a few weeks earlier.

Jack was busy adjusting two massive Pennzoil signs against a partition. When he spied us he shuffled in our direction in his loose fitting sandals. "Ah, Nadia. It's so nice to see you."

What am I? Chopped liver?

"It's nice to see you too, Jack." Nadia purred.

"Can I help you find anything?" he asked, not bothering to pull his eyes away from her.

I cleared my throat. "How much are the vases?"

He spared me a glance. "The ones in the two-dollar bin?"

I shot him a sarcastic grin.

"Come here, Nadia. I have something I want to show you." He led her into the back room and I curiously followed behind. Boxes were stacked on top of boxes. There was so much stuff that I couldn't tell what anything was. He stopped suddenly and pointed to a very large birdcage resting in the corner. "Do you think Bing would like this cage? I realize it's rather large." His eyes sparkled with excitement. "When I saw it I just had to buy it for you. I mean I had to buy it for Bing."

"It's beautiful, Jack." Nadia turned to face him. "Bing would love the space. Thank you for thinking of us."

"I think about you all the time," said Jack.

Nadia blushed.

I rolled my eyes, feeling like a third wheel.

"I'll bring it by your house later, if you'd like," he suggested.

"That would be lovely."

"I also have a side-table that I figured you might be able to spruce up by painting it. You know, like the one you have in your living room," Jack suggested.

"I'll give it a try," Nadia agreed.

Since I was obviously being excluded from their conversation, I turned around and headed back to the main area of the store and tugged two vases from the bin. Plopping a five-dollar bill onto the counter I called out, "I'll be waiting for you in the car, Nadia, and Jack, give Nadia my change."

I actually heard them giggling as I meandered toward the door.

I was glad I had thought to bring Mark Twain's *A Tramp Aboard* with me because it was several very long moments before Nadia slid into the passenger's seat. I was laughing out loud at Twain's good humor as he absorbed the scenery in Europe.

"What are you laughing about?" she asked.

"This book." I placed it back on the console.

"Good. I assumed you might be laughing about me and Jack."

"Never," I assured. "I would never laugh at two people falling in love."

"Love? Don't be silly." Nadia tsk-tsked.

"Whatever..."

She handed me my change and twisted the knob on the radio down a notch. "Turn Your Love Around," by George Benson was blasting from the stereo speakers.

"What? Don't you like George Benson?" I asked.

"I love him," Nadia assured. "It's just that I wanna tell ya about moving from Benwood to Charleston on our drive home."

"Perfect!" I pointed my finger at her. "Even better than listening to George or reading Mark Twain."

A New Life

Charleston, West Virginia
July 21, 1924

The house overlooking the Kanawha River was two stories, with a blue frame, and a wide porch and two turrets. To Nadia, it looked like a mansion—stately and welcoming. She, Isaac, and William eased from the cab of his truck and stood staring at it for a long moment. "Well Nadia, what do you think?"

"It's beautiful," she replied.

"It's big," William exclaimed.

Isaac laughed. "It's not all ours. I've been renting half the house from Mrs. Carrington who lives here too. It was converted into two areas about three years ago." He pointed to the left hand portion of the building. "This side is ours."

They walked up the steps to the broad porch and Isaac inserted his key into the door. "Shall we?" he asked, before sweeping Nadia into his arms. He carried her across the threshold while William guffawed behind them. The hardwood floors were worn and battered, showing signs of hard use, and the fireplace in the front room was blackened with years of smoke. However, the light was nice and the ceilings were high. A narrow staircase led upstairs to a hallway that extended in both directions. They passed a small bathroom before entering the kitchen, where Isaac made a swooshing motion with his hand. "What do you think?"

"It's amazing, Isaac," was all Nadia was able to say. She never dreamed *she* would actually live in her own home that was so splendid—rented or not.

"C'mon, William. Let's unpack the truck and get you two settled in."

Before they had wed, back in Benwood, Nadia and Isaac had inventoried the old shack up Flanner Holler and Nadia was thrilled when Isaac suggested they pack some of the contents to bring with them to Charleston. She had always considered the secondhand furniture, and handmade curtains to be bland and without value. But in the end, they had hauled a small kitchen table, a bench her daddy had carved, an aged clock, three sets of curtains, two cast iron skillets, her daddy's Bible, and of course, a suitcase full of clothes from Benwood, some one hundred fifty miles south, to Charleston.

Nadia also had a letter, written by Mr. Bentz, to deliver to Mr. Geary of The Diamond Shoe and Garment Company on Capitol Street. The correspondence was a recommendation, which Nadia hoped would help her secure a job at the department store, and if all went well, she would call on Mr. Geary the following day. Nadia was proud in that way. She didn't want Isaac to think she was like her mama, Rebecca, simply trying to find a free ride through life. She wasn't afraid of hard work.

While Isaac and William were making the numerous trips from the truck to the entryway of the duplex, their neighbor, Mrs. Carrington, came over to introduce herself. "Hello, Nadia. I'm Mrs. Carrington, but you can call me Ruth."

Nadia was greeted with a waft of warm, apple pie scent lingering in the air. "It's a pleasure to meet you, ma'am."

"I've brought you over some stew and apple pie for your supper tonight. I figured you be too busy to cook, since you are newlyweds." Ruth's hand flung over her face as if to hide the dash of delight coming out, and the only way Nadia knew she was smiling was because her puffy cheeks elevated her spectacles. The older woman must have noticed Nadia's face flush with embarrassment, because she quickly added, "And moving is quite exhausting."

Nadia released the breath that had caught in her throat. "Thank you. You are very kind." She stared from her halo of white hair, to her bright-eyed face, to the over-the-top string of beads hanging around her neck then to the apple pie and bowl of stew the other woman was still holding in her hands. "Please come in."

Ruth, knowing her way to the kitchen, darted toward the envelope of a kitchen and placed the goodies on the table. She dropped down into a chair and motioned for Nadia to join her. "Who is the young man that is helping you move?"

"He's my brother. His name is William. He's going to be living with us."

"Oh, I see." She peeked at Nadia. "Are your parents okay with him living with you?"

Nadia briefly measured her response. "My daddy died a few years back and my mother is... she left us."

"She left you?" Her eyes popped open.

Nadia felt her cheeks heat up. "Yeah, I've been taking care of William for a long time now. It was just us until I met Isaac."

"Isaac seems to be a fine young man. I think you caught yourself a good one. I know he travels a great deal with his work, so if you ever need anything or get bored staying around the house, always remember that I'm right next door."

"Thank ya." Nadia nodded. "I'm hoping to get a job at The Diamond Shoe and Garment Company. I am gonna go visit Mr. Geary tomorrow, and Isaac got William a job working at the newspaper in the printing department. So, hopefully we won't get too bored while Isaac is away."

Ruth laced her fingers through the necklace draping her neck. "Good for you! You'll be just fine living here and I always cook too much food so you can count on me to fix you up some supper or care for your little ones once they come along." She added a bright, reassuring smile.

Nadia realized that she seemed like a master at sounding optimistic and cheerful. "Thanks, Mrs. Carrington."

"Please call me Ruth since we're going to be neighbors."

"Okay, Ruth."

"Are you a good cook?" she asked.

"Not really," Nadia responded. "However, Mr. and Mrs. Bentz were very kind to us after my daddy died and Mrs. Bentz taught me a little about cooking. She was the kind of woman who believed that one should put their hands into their prayers so she often cooked for people in our community."

"Really?" Ruth's eyes lit up. "I have a cookbook that I'll share with you. It's not organized by recipes as much as it is by people."

"Organized by people?" Nadia pondered. "I don't understand."

"Ah, it is. My cookbook has folk's names in it, so that if they need comfort I can easily find their favorite dish. For example, Mr. Clark's favorite dish is rhubarb pie, so when he is feeling poorly I'll bake a pie for him." She glanced over the sparkling eyeglasses perched on the end of her nose. "His wife died a few years ago and he has been suffering from arthritis." She shook her head sadly. "Some days he can hardly get out of bed. Then there is Mrs. Pawley who adores chocolate chip cookies. She has cancer, so I try to visit with her once a week and take her a warm plate piled full of them."

Nadia stared at the other woman with amazement. "That is the sweetest thing I've ever heard! I do want to see your cookbook." She leaned in close and asked, "So, just between you and me, what's Isaac's favorite?"

Ruth confided, "Apple pie, of course."

Nadia glanced down at the table. "Naturally. So, you will teach me how to bake, right?"

"Young lady, I'll teach you everything I know about cooking and about men, too."

Nadia laughed. "Thanks."

"Thank you. I'm looking forward to spending time together." Ruth rose from her chair and fluffed her hair. "Well, I best be moseying on. I want to stop in and check on Mr. Clark, he's a bachelor. Anyway, ya'll have fun tonight and rest assured the walls are thick in this house. I can't hear anything going on over here no matter how hard I try." She offered a cunning wink before heading out the door.

∞∞∞∞

The following morning, Nadia drove Isaac and William to work at the *Charleston Gazette*, located on Hale Street.

"Good luck, baby." Isaac kissed her on the cheek. "You look beautiful today and you are going to get this job."

"I hope so. I'm a little nervous."

"Don't be. You'll do great. Come back to pick us up around four o'clock."

She squeezed his hand. "I'll be waiting right here."

Nadia followed the directions that Isaac had written down and easily found her way to the Diamond Shoe and Garment Company, squared her tiny shoulders and pushed through the door. She nervously scanned the first floor before stopping at the cashier's station in the men's shoe department. "Excuse me," she cleared her throat, "could you tell me where to find Mr. Geary?"

The woman smiled delightfully. "Yes, ma'am. His office is the last door on the right." She pointed toward the back of the store.

"Thank you," Nadia replied. "Should I stop at his secretary's desk and introduce myself?"

"Nah," the woman, whose nametag indicated her name was Edna, leaned in and lowered her voice. "He doesn't have a secretary, per se, and he is very approachable. Just knock on the door." She arched a brow. "Are you here to inquire about work?"

"Yes." Nadia wrung her hands nervously. "I just moved here from Benwood and I'm really hoping he'll have something available."

Edna winked. "Good luck. This is a great place to work and Mr. Geary is a very kind man."

Nadia strolled through the ladies' fine apparel section and stopped in front of the closed door. She raised her hand to knock then lowered it again. She could feel her stomach beginning to knot up. Taking a deep breath and exhaling slowly, she timidly rapped twice. She prayed he wouldn't want to discuss her limited credentials.

"C'mon in. It's open."

She twisted the knob and slowly approached the man seated behind a large mahogany desk. "Hello, Mr. Geary. My name is Nadia and I've come to inquire about work." She handed the letter of recommendation across his desk.

"Ah, Nadia. It's a pleasure to meet you. Mr. Bentz told me you'd be stopping by. Please have a seat." He motioned to a chair sitting opposite his desk and slid the letter from the envelope. Nadia watched as he slowly read the note. "Very impressive. Mr. Bentz indicated that you are the best worker he has ever employed and that you are an expert at creating displays."

Nadia blushed. "I wouldn't say I was an expert, but I did experiment quite a bit when I was working for Mr. Bentz." She swallowed deeply. "The dry goods store in Cooey-Bentz Building is not as elaborate as The Diamond Shoe and Garment Company though."

"Yes, I know. I've been in Benwood several times over the years and know Mr. Bentz quite well. He is a fine man." Mr. Geary placed the letter on the desk. "So, do you mind to show me your presentation skills?"

Nadia gulped. "Right now?"

"Yes, please."

"Okay." She smiled, trying to remain calm. "Where would you like me to start?"

He rubbed his chin thoughtfully. "Follow me and I'll show you around the store. I'm sure something will catch my eye." He indicated she should join him.

As they strode through the various departments, Mr. Geary paused to introduce Nadia to the employees in each department. He paused at the desk where the woman, Edna, whom she had met earlier, was working. "Edna, this is Nadia and she is applying for a job. Her reference spoke highly of her ability to create appealing displays. Could you show her to the new stock of dress shoes and where we keep the store display fixtures?"

"Yes sir," Edna replied. "Please follow me, Nadia."

Nadia followed Edna to a room located just behind a bracket presenting stylish shoes for customers to consider.

"Here are the shoes, and the shelving systems, clothes racks, which you won't need for shoes, over there are cubes that can be stacked and cloth to cover them," Edna explained as she pointed to various areas in the storage room.

"I'm really nervous," Nadia confessed. "The displays I made at the store where I worked in Benwood were made from left over shipping crates and scraps of material."

"Honey, don't you worry. I'll help you and I'm sure it will be great," Edna said encouragingly. "Where do you want to start?"

Nadia shrugged. "I guess I should see the shoes first."

"Right here." Edna bent over and opened one of the boxes.

Nadia stared at the brown and black cap toe oxfords and two-tone sport shoes and a hint of smile crossed her lips when she recalled the shoe Isaac had lost in the river. *I can do this.*

About an hour later, Mr. Geary stopped by and stared at the display for a long moment. A wide smile crept across his lips. "Can you start tomorrow?"

"Yes, sir."

"Welcome aboard." He turned and shook her hand tenderly.

Edna gave her a wink of approval. "I'm looking forward to working with you, Nadia."

Nadia was walking on cloud nine when she left the store. She had started a new life. She didn't have to walk or ride a bike up a muddy holler to get to work. She was married to a kind man, William was being cared for, and she just got a job in the big town of Charleston. *I'm as lucky as a baby calf in a clover field.*

She decided to treat herself to a walk down Kanawha Street to window shop and to learn her way around town. She passed The Five and Dime Store and Scotts Drug Store. There were signs everywhere: Beauty Shop, Bookstore, Have Your Suits Pressed Here, Drink Coca Cola. If you could name it, Nadia could see a sign advertising it. She was about to approach a stand where the vendor sold fresh fruit and vegetables near the corner of Hale and Kanawha Streets, when she noticed a Negro woman, dressed in what appeared to be a housekeeper's uniform, attempting to buy an apple.

"Apples are ten cents," she overheard the dealer say to the woman.

"But the sign says they are two cents," the woman pointed out.

"They're two cents for white folks." He tilted his head and added a smirk. "And don't touch 'em. Nobody would want to buy one after you put *your* hands on it."

Nadia thought about the immigrants she knew in Benwood, and felt her blood start to boil. She decided to intervene. "That ain't fair," she alleged.

"Don't..." the woman lowered her eyes and shook her head.

The portly man addressed Nadia. "What ain't fair?"

"That you're trying to overcharge her," she replied diplomatically.

His beady eyes narrowed. "She's a nigga."

His mouth sure ain't no prayer book, Nadia considered. She knew the Golden Rule applies to everyone and that every person should be treated with respect—no matter where he or she comes from, how plain their clothes were, or the color of their skin. "Fine." Nadia replicated a smile. "I'd like to buy a two cent apple." She pulled two pennies from her change purse and handed it to the man, picked up the biggest, juiciest apple she could find and turned to place it in the other woman's hand.

"So," he drawled, "you're a nigga lover."

"Yep. As I have just witnessed, the color of one's skin has nothin' to do with the amount of integrity they possess."

"Integrity?" The man winced.

For a split second, Nadia thought he looked confused.

"Look lady, don't come back to my stand again to buy nothin'. You ain't welcome here." His chubby chin jiggled when he flashed a wide smile exposing his yellow-tainted teeth.

"Oh, believe me, I will never purchase anything from you." Her eyes sparkled with an ill-felt confidence. "I'll spread the word, too."

The man offered a very offensive jerk of his hand, causing Nadia to roll her eyes dismissively and turn to walk away.

"You didn't need to do that for me," the woman whispered.

"Yes, ma'am. I did."

The woman attempted to hand two pennies to Nadia, but she brushed them away. "You keep your money. It's my treat. There ain't no reason for folks to be bigots. It only shows their stupidity and ignorance."

The woman offered a sad smile. "Well, thank ya. I appreciate it," she said, before scurrying down Hale Street. Nadia watched the other woman for a long time, until she disappeared around the bend in the road, wondering what it would be like to be a black woman in this prejudiced world. *I hope I live long enough to see the change—and maybe even be a part of it.*

Charleston, West Virginia

1982

"Wow, Nadia. You were brave!" I said.

"No Dee, I was scared to death," she admitted. "My voice was quivering the whole time I was speaking to the man."

"You were," I insisted, "being brave is when you do things *because* you are afraid. That's what bravery is to me."

"Okay," she smiled, "maybe I was a little brave. A Southern lady always has strength in times of trials and is a source of joy in good times."

I nudged her. "Nadia, you are the epitome of a Southern woman."

She blushed in return.

"Plus," I continued, "you were really blessed to find a job so quickly, and Mrs. Carrington sounds like she was a charmer."

"I was very fortunate to get a job. I worked all the way up until our first daughter, Sue, was born, and Ruth was wonderful. She kinda became like a second mother—much like Mrs. Bentz." She pointed to her front door. "Dee, do you have time to come in?"

I stole a look at my watch. "Sure, I have a little over an hour before I have to go anywhere."

"Me too. Since I cancelled my dinner date with Sylvester Stallone I have nothing to do again tonight, and Tom Cruise heard I was busy so he hasn't called lately."

"I thought Jack was stopping by later," I reminded her.

She jiggled her eyes mischievously. "So many men and so little time."

I laughed out loud. "I sure wish I had that problem."

"You will," she assured me. "As soon as you quit hanging around with old folks."

I tilted my head. "What *exactly* do you mean by that statement? I have a very, okay somewhat, interesting life."

"I know you do, Dee." She motioned for me to follow her into the living room.

I dropped down onto the couch, slung my shoes off, and tucked my feet underneath me, while Nadia snuggled in close and opened her memory album.

"I just wanted to show you a few more photographs from my 1924 album. I've been spending quite a bit of time thinking about Benwood and moving to Charleston ever since I started telling you the stories about the mining disaster and Isaac."

"That sounds good." I peeped over her shoulder.

"Here is a photo of Mrs. Ruth Carrington. Isaac took a picture of her." She pointed toward a specific photograph taped on the page.

"She looks a bit eccentric with those multiple strings of beads hanging around her neck."

"She was." Nadia seemed to get lost in her memories for a moment. "The day I got my job and came home and told her about the unfortunate incident at the produce stand she, in turn, told me a story. It was quite a frightening tale and one that I'll never forget."

"Really?"

"Mmm, hmm. Would ya like to hear it?"

"Sure."

Bigots and Cowards

Charleston, West Virginia
July 22, 1924

N adia was still fuming over the vendor's absurdity when she pulled in the driveway of her new home on Kanawha Boulevard. She noticed Ruth outside by the clothesline. As she walked over to her she could smell a pop of freshly washed sheets as they snapped in the summer air.

"How is the honeymoon going?" her neighbor asked.

"I can't complain." Nadia flushed.

Her friend arched a brow of approval. "Excellent. Did you get the job?" the bubbly woman inquired.

"I sure enough did." Nadia replied. "Unfortunately, I also had a disturbing incident occur while I was in town."

Her hand rose to cover her mouth. "Is that so? What happened?"

Nadia recapped the confrontation, careful not to leave out any details.

Ruth shook her head despondently. "Southern girls are raised to think whatever they want, but to choose the words that escape their mouths very carefully." Ruth waved a bejeweled hand. "Sometimes you find yourself in a place where it is safer to keep your mouth shut."

"Perhaps," Nadia skeptically murmured.

The older woman sighed. "Would you care for a glass of sweet tea?"

"That would be wonderful."

Nadia followed her into the kitchen and helped her fold the clothes she had pulled fresh off the line. She watched as the woman carried a bushel of peas waiting to be shelled into the pantry. She then poured two glasses of sweet tea and dropped a slice of lemon into each tumbler. Ruth held her glass in the air. "Even though we're living in these dreadful days of prohibition and can't add a shot of vodka to our tea, a toast is in order. May I?"

"Absolutely."

"Here's to friends, old and new."

Nadia maintained eye contact as the glasses made a perfect ring, clearly buying into the old tradition that it was bad luck not to do so.

"So Nadia, I want to tell you a story." Ruth Carrington began, "Some eight years ago, in 1916, my husband and I were staying in Abbeville, South Carolina. My husband, Ben, and I had been living there for less than a month. We had gone to settle my parent's estate after my father passed away. One day we were playing a game of Legan chess, which differs from standard chess by the starting positions of the pieces on the board. Anyway, Ben was winning because Southern ladies always let their men win, no matter what," she added with a teasing grin. "Suddenly we heard quite a commotion going on outside on the street. Ben stepped outside to see a man named Anthony Crawford, was being beaten as they dragged him through the street."

"Who was Anthony Crawford?" Nadia asked.

"He was a successful landowner and farmer. My parents knew the family very well."

"Why were they beating him?"

"He was a black man and he had allegedly cursed at a white man for offering him a low price for the cotton seed he was trying to sell. The white man said he was too rich for a Negro."

"That's horrible."

"It was indeed." Ruth twirled the string of beads adorning her neckline through her hand as she drifted back to the dreadful day. "About three hundred, or so, white people wanted to show other Negros what would happen to them if they got *insolent*."

Nadia's eyes grew wide.

"I begged Ben to go outside and try to help him, but he insisted there was nothing we could do. If we attempted to interfere we could possibly be treated the same way. 'What about his wife and children, Ben?' I pleaded. 'Please let's do something!'"

"Did you try to help him?"

A tear formed in Ruth's eye. "No. We cowered in the safety of our own house and wept. We didn't even try to help."

"What ultimately happened to him?"

A gloomy expression crossed her face. "Well, the beating took place for hours. Eventually, they took him to the county fair grounds and strung him up to a tree and riddled him with bullets. I don't know what the mob did with the body afterwards because they never found it."

Nadia digested this for a moment.

Ruth explained, "The town folk ordered his family to vacate their land, wind up their business and get out of town. So, of course they left."

"Did you stay there after the incident?"

"Absolutely not. We left as soon as we could and came back to Charleston. I couldn't stand looking at the people in Abbeville. They made my stomach turn and the bile rise up in my throat every time I recalled what they had done to Mr. Crawford, and believe me, I'll never go back."

"I'm sure it was very disturbing," Nadia empathized.

"I told you this story because you need to understand that even though you're white, you're still a woman, and bad things happen to women," she nodded to emphasize her point. "Heck, they can happen to anyone who gets involved trying to stop hate crimes. Even in this day and age, racial stereotypes are pervasive and not limited to the Deep South."

Charleston, West Virginia

1982

"Wow, a lynching. That's appalling," I acknowledged.

Nadia attempted to explain. "I feel Ruth shared the story with me so I would be careful. Those times were tough and people were hateful and ignorant."

I nodded, knowing this much was true.

"Dee, did you know that even in 1924 the Ku Klux Klan had members in several locations in West Virginia. They opposed everyone. The KKK was against Catholics, Jews, immigrants, and labor unions as well as folks of color. They even managed to have several Catholic public school teachers fired. Isaac told me they dabbled in politics and influenced the outcome of a number of local elections."

"That's sad."

"Yep. When I told Isaac what happened with the vendor at the fruit stand, he reminded me of how violent people can act and told me I best be careful."

"Did you ever see that woman again?"

"The lady I met at the vendor's stand?"

"Mmm, hmm."

Nadia shook her head. "No. I never did. I always kept an eye out for her when I was heading to or home from work but I never saw her again."

"Why did you look for her?" I probed.

"I wanted to invite her to have lunch with me," she replied, matter-of-factly. "I was raised in Benwood, and we were all immigrants, so to speak." She shrugged. "I also wanted to prove a point to folks. I wanted people to understand that friends can come from all walks of life. I mean, just look at this photograph of some of the miners who died in the Benwood mine." She pointed to a specific picture. "Do you think the Serbs, Greeks or Slovakians cared who was working beside them? Of course they didn't. They were neighbors and they took care of one another."

"Well," I sighed, "I'm glad things are finally changing."

"I helped make it change," Nadia almost boasted.

"Really? How so?"

"Well," Nadia bowed her head shyly, "Isaac and I were there when they integrated the schools in West Virginia. It's not a huge deal but every act of bravery and kindness counts, right?"

"Absolutely. Do you mean desegregation? Do you have any pictures?"

"No. Although Isaac did write some articles about it." She tapped a finger to her temple. "But I remember it well. It's all up here."

I pulled the blanket over my legs, to escape the freezing air circulating in the room, leaned back on her comfy sofa, and prompted, "Then tell me a story."

Taking a Stance

Greenbrier County, West Virginia
September 19, 1954

"Would you like to drive down to Greenbrier County with me this morning?" Isaac asked, placing his breakfast dishes into the sink.

"Why are you going down there?" Nadia asked, plunging her hands into the warm soapy water.

"I'm covering school integration throughout the state and there are public protests going on at White Sulphur Springs High School."

"I thought the Supreme Court ruled that *separate but equal* has no place in our society and that divided educational facilities were unequal."

"They did," Isaac responded. "Nonetheless, the protests are delaying integration in Greenbrier County and I was assigned to cover the story for the newspaper."

"Okay, I'll go with you." She plucked the plug from the sink and rinsed out the remaining suds. "Let me change clothes and run a brush through my hair."

The drive took a little over two hours and the scenery was splendid. Fall in West Virginia is almost unimaginable unless one has truly seen it in color. The ridge crest sparkles with leaves turning rich hues of stark reds, buttery yellows, and shades of auburn and on the distant mountains, forests of dark evergreens stand strong, while shadows of clouds glide slowly up the valleys. At every turn in the road, Nadia was awed anew at God's creations, and she was slightly disappointed when the picturesque drive came to an end.

When they arrived she saw crudely painted signs that read, "No Negros Wanted in Our Schools," and some three hundred white parents and students picketing. Voices were shouting, threatening bodily harm to the Negro students who were standing at the opposite end of the field, and several dozen folks stood in the background ogling. Nadia watched as Isaac took a few snapshots before he zigzagged through the crowd over to where the superintendent refused to interfere.

"What's going on here, sir?" Isaac asked. He pulled out a notebook and ink pen.

"At the present time boards of education are free to act, and I shall take no action."

"You are aware that on May 17th the United States Supreme Court ruled unanimously that the doctrine of separate but equal has no place in our country."

"Of course, I'm aware."

"Then why aren't you doing anything?" Isaac further probed.

"Because integration has been left in the hands of the *county* boards of education." He shrugged. "The Greenbrier County Board of Education called off integration and ordered all students back to the schools they attended last year."

Isaac took a photograph of him. "I'm a reporter for the *Charleston Gazette* and will be covering the incidents that occur here today."

The superintendent lowered his brow. "I don't care who you are or what you write."

"Perfect." Isaac pointed at him. "That's how I'll start my article, with your insightful quote," he sarcastically countered.

"Wait." The man reached out and grasped his arm. "You have to understand that the white citizens who live in this area weren't asked whether they wanted segregation in the schools or not. They weren't even given a chance to vote on this issue."

"Nonetheless, you are engaging in open and flagrant violations of the laws of the state," Isaac reminded him.

From a distance, Nadia watched in disbelief as folks started shouting obscenities at the students:

"If God meant for the races to mix, He would
have made us all one color!"

"If you start puttin' them into our schools, our children are gonna think it's alright to marry them!"

"We're not going to stand for this!"

"There will be bloodshed!"

Soon there were eggs and apples being thrown at the black students and Nadia wondered where the police officers were. She took in a deep breath, squared her shoulders and sauntered from her safety, as a bystander, over to where the Negro students were huddled. She planted her petite little body directly in front of them and watched for objects being thrown in her direction in case she needed to dodge them.

Isaac saw her and scampered over to where she was standing defiantly. Her arms were crossed, and her chin held high. "What are you doing, Nadia? You're going to get hurt."

"I'm taking a stance, Isaac. Would you please stand beside me and take pictures of those stupid folks." She pointed toward the protesters. He started snapping away. Apparently, the notion of having their photographs taken thwarted their actions and the hum in the crowd lowered to a whisper. Much to Nadia's disbelief, other white folks, who were simply watching the incident take place, began walking in her direction. One by one, men, women and about one hundred white students gathered their courage and soon a barrier of bodies was protecting the Negro students.

The standoff lasted for over an hour before the protesters

dissipated, and seemingly, due to the white folks finally stand-ing up to injustice, along with Isaac's article in the newspaper, White Sulphur Springs High School quickly became integrated and both races became determined to work together for better education for all the children of all the people.

Charleston, West Virginia

1982

Nadia glanced at the clock. "Oh Dee, I'm so sorry. I've been talking too much and I know you have things ya need to tend to."

"I'm fine," I replied. "I'm very impressed that you had the guts to stand up for the students."

"Well, thank you. I will admit I was quite proud of my actions. I haven't made a big difference in this world, but I'm as pleased as punch about the few things I've done to help other people." She patted my arm. "I do realize that Negro is not proper terminology in today's society, but that is how it was back then."

"I understand," I replied. "Nadia, you have led a remarkable life, whether you realize it or not."

She smiled modestly. "Do you have time to look at the rest of the pictures? I only have a few more photos to show you."

"Sure."

Happy as a lark, Nadia turned to the last page of her beloved memory album. "Here's what William looked like back then."

"Is William still living?"

"Yeah, he's doing fine. He grew up to be a respectable man and has three children. They're grown now, of course, but William and his wife, Gail, live in California. He retired from the post office a few years back." She turned her attention back to the page. "This is the only picture I have of my daddy," Nadia explained, placing the album on my lap.

Her daddy stood proud and tall, and even though the picture wasn't clear I could determine some of his facial features, which included prominent cheekbones, and broad chin. He was wearing an undershirt and stained trousers. He put me in mind of an American Indian, though I couldn't determine the exact hue of his skin through the shades of gray.

"Will you show me the picture of you and Isaac on your wedding day again? Since I was driving I couldn't get a good look at it."

"Sure, let me grab my pocketbook." She fumbled around to find her wallet and finally presented me with the photo I had only caught a glimpse of earlier.

"Here ya go." She grinned. "We were just two young kids doin' the best we could."

I squinted my eyes. "Wow, you really were just a kid—a tiny little thing. What did you weigh? Like, ninety pounds soaking wet?"

She laughed under her breath. "I wasn't suffering from malnutrition if that's what you are asking. William and I ate quite a few meals at the Bentzes' house, and most of the time I would bring her a basket of eggs, because that's about all we had to offer. I raised chickens and always had a plentiful supply of eggs in the coop." She paused as a reminiscent smile wrapped up around her cheeks. "Almost every time we would go to their house I would take eggs, because that's really all we had to offer. Nonetheless, Mrs. Bentz would always act as though it was the best thing I could have given to her. It's funny how kind folks just know how to make people feel important, isn't it, Dee?"

"Yes, for sure. It's what we are meant to do in this life—make other people feel important," I replied honestly. "Most the time I hear folks trying to beat others down, and we have plenty enough of that in the world, don't you think?"

"Absolutely!" Suddenly, something on the page of her memory album caught her eye. "I have one other photograph of me that Isaac took before we were married. Take a gander at this."

I stared at the beautiful young woman in the black and white photo. She was walking away from the camera, somewhere in a lush, overgrown field. The longer I examined the picture, the more the colors came alive. I could see the abundant olive-green trees in the background. I could make out the pleats in her auburn skirt and the cream-colored ribbon attached to her straw hat—if only in my mind's eye.

"You were lovely, Nadia."

She smiled unassumingly. "So Dee, those are my memories, right here in black and white. I'm so thankful I still have these pictures. These snapshots are proof that once, even if just for a heartbeat, everything was real and sometimes even perfect."

"Thank you for sharing your stories with me, Nadia, they are remarkable tales. Do you have any more photographs tucked away somewhere?" I asked.

"No. My other albums were ruined in the flood of '36. Thankfully, my memory album from 1924 survived. Although, I recently ran across a box of old articles Isaac had written over the years and I want to give 'em to you to sort through. You might find some interesting ideas to write about."

"I'd love to read them."

Nadia gingerly closed her memory album and tapped it with her hand. Her eyes sparkled dreamily and a grin tinted her rosy cheeks. "You should have seen us in color."

∞∞∞∞

Later that evening, I rapped on the back door one time before twisting the knob and pushing it open. I immediately observed Jack and Nadia holding one another close and it looked as though they were… hum, interesting.

"Are you guys canoodling…?" I stammered. "I mean dancing?" My eyes shifted to the large bouquet of lavender flowers situated in a vase on the kitchen table, down to where Max was sleeping underneath the table, over to the bottle of wine sitting near the sink, then back to Nadia and Jack. They casually pulled away from each other.

"Yes Dee," Nadia confirmed, as she pressed the imaginary creases from her dress. "We were dancing."

"Oh… I don't hear any music." I shifted my weight awkwardly.

Jack impulsively reminded me of his beloved Nietzsche quote. "And those who were seen dancing were thought to be insane by those who could not hear the music."

I offered him a half smile and turned up my ear. That's when I perceived the sound—music resonating from the living room. Bing was enticing them with his whistles, trills and gurgles. It was truly symphonic. "It must be mating season," I suddenly blurted.

The astonished expressions on their faces did not go unnoticed.

"I meant because Bing was singing." I nodded to affirm my previous statement. "Yep." I started babbling uncomfortably, "Bing is putting on a show." I clarified, "It must be mating season for Bing."

I noticed Nadia's shoulders relax slightly.

"Well," I faked a yawn, "I best be getting home and leave you two to… your frolicking… or whatever it was you were doing when I burst in. Sorry." I pointed to the kitchen door. "Yes, I'll just mosey on home now because Gabby and I have big plans for this evening." *I have big plans with my cat? Geez. I sound like a total loser.* "I just stopped by for a second to check on you."

"Okay. Here, Dee." Nadia hurriedly shoved a small box into my arms. "These are all of Isaac's newspaper articles. I thought you might want to look through them sometime since you enjoyed looking at the old photographs." She gently grasped my elbow to escort me out. "I do hope you and Gabby have a fine time tonight."

As she was attempting to heave me out the door, I leaned in

and whispered, "A lady always wears clean undies…"

"Or, none at all," she interrupted me brusquely; a devilish grin traversed her face.

"Nadia!" My jaw dropped open in mock repulsion. "You are downright scandalous!"

Holding the cedar box tightly in one hand, I turned and bounced down the steps—chuckling heartedly when I heard the deadbolt snap shut behind me.

$$\infty\infty\infty$$

I woke up with Gabby's raspy tongue exfoliating my cheek. "Just five more minutes, Gabby. Please?"

She gave me about a ten-second reprieve and started in again. A soft purr echoed in my ear, causing me to open one eye and look at the clock on the bed stand.

"I can't believe I've slept this late, Gabby. Why didn't you wake me up?"

She gave me a disgusted look, swished her tale in my face and gracefully leapt from the bed. I followed her into the kitchen, started the coffee brewing and thrust a can of Posh Banquet into the can opener. She watched in anticipation, as it twirled around in circles, and reluctantly jumped from the countertop when I placed her breakfast on the floor.

After two cups of caffeine, I slid into my Lee jeans and plain red T-shirt, ran a brush through my hair, picked up the box that Nadia had given me and rushed out of the house.

"Nadia!" I pounded on my neighbor's door. Bing greeted me with a "Good Morning" chirp before Nadia welcomed me enthusiastically.

"Good morning, Dee. What are ya doin' here so early in the morning?"

"I was sorting through the cedar box stuffed full of newspaper articles written by Isaac last night. I want to ask you some questions."

"Sure." She motioned for me to follow her. "Would ya like a cup of coffee and a powdered donut?"

"Yes, thank you. I love powdered donuts."

"Have a seat," she said over her shoulder.

"Hey, did you have a nice date with Jack last night?" I asked.

"I did, Dee. I haven't danced in years."

"Good for you."

Nadia glanced at me. "What were you wondering about the newspaper articles?"

"What kind of stories did Isaac write?" I asked.

"He wrote about everything. He covered mining disasters, like the Everttville explosion in '27, the Pond Creek explosion in '40 that killed ninety-one men, and the last one he covered, I believe, was the explosion in Farmington where seventy-eight men died. That was back in '68 and right before he retired. He went to roof falls, mine fires and explosions, like what happened in Benwood, dozens of times."

I plopped the cedar box on the top of her chrome-and-vinyl dinette table and lowered myself into a small chair.

"Depending upon what was going on at the time I would sometimes tag along with him. Soon after I moved to Charleston I spent the weekends with him when he was writing about the thirty-fifth presidential campaign. John Davis, from West Virginia, was running against Calvin Coolidge and we spent a great deal of time covering the election."

"What else did he write about?"

She seemed to consider this carefully. "Ah, he spent years writing about unsolved mysteries."

My interest piqued. "Did you say unsolved mysteries?"

She nodded. "Mmm, hmm."

"Like what?"

Nadia paused to remember. "Probably the most famous story was the disappearance of the Sodder children. Actually, I accompanied him when he covered the Sodder story for the newspaper, because he got a call and was asked to go to Fayetteville on Christmas Day and I insisted I go with him, because I didn't want him to be alone." She fluttered her hand in the air. "Anyway, it was one of the oddest things I'd ever seen."

"I was trying to make heads or tails out of the Sodder family articles I found in the box last night. What really happened?"

"Well… let me think. After Isaac got the unexpected telephone call, we jumped in the car and took Sue and Sophia over to Mrs. Carrington's house." She looked up at me. "We had already moved from the duplex we rented from Mrs. Carrington into this house by then and the girls were teenagers at the time. They insisted they'd rather visit with her than spend the morning driving to Fayetteville, and Isaac agreed due to the nature of the incident he would be uncovering. Sue and Sophia adored Ruth Carrington because she occasionally took care of them when they were children." She shrugged her shoulders. "You know how teenagers are. They don't want to spend time with their parents."

Some adults don't either. I reflected on how seldom Sue and Sophia visited Nadia and wondered if she missed spending time with her daughters. "What was the 'nature of the incident'?" I asked.

Nadia handed me a cup of steaming coffee before she rummaged through the box to find the article with a photograph of a smoldering house. "I believe surviving a tragedy is a very personal event, and sometimes there are no words to describe what one feels while fumbling through it. And all the time, everyone else just goes about with their own lives—sympathizing for only a moment on the heartbreak of others—or perhaps not showing any compassion at all." She shook her head. "I just don't understand it." She handed me the newspaper clipping and I gazed at the black and white photograph of the remnants of a house with timber still seething—tendrils of smoke evaporating into the air. I had noticed the photo the night before, but even after searching through dozens of articles I didn't discover the outcome of the case.

"So Nadia, tell me the story about the Sodder children," I requested, taking a big bite from a powdered donut.

She topped off my coffee. "It occurred almost forty years ago, but I'll tell ya everything I can remember."

Up in Smoke

Fayetteville, West Virginia
December 25, 1945

I t wasn't a normal Christmas Day with roasted turkey, fruit-cake, caroling and opening gifts under a brightly decorated tree. Neither for Isaac and Nadia nor for a family in Fayetteville, whose home went up in flames the night before.

"So Isaac, why didn't you want Sue and Sophia to come with us this time?"

Isaac made a sharp turn leading to Route 16. "I have no idea what we are going to see when we get to the house, but when Mr. Jackson called from the *Charleston Gazette*, he indicated it was very important that we have something about the fire in the newspaper tomorrow."

"There are always house fires, Isaac. What makes this one so important?" Nadia dropped her knitting needle into her bag.

"They are assuming several children died in the fire," he replied regretfully.

"Several children!" Nadia gasped. "That's horrible. Those poor parents."

"According to what Mr. Jackson knew when he called this morning, a fire broke out at around one o'clock in the morning and the house went up in flames in no time. He believes that the parents and some of the children escaped and when they called the Fayetteville Fire Department they couldn't get an operator response. A neighbor made a call from a nearby tavern, but still no one answered the call, so he drove into town and tracked down the fire chief. They performed the usual method of a fire alarm, which is really a phone tree system whereby one firefighter telephones another and so on. Even though the fire department was only a couple miles away the crew didn't arrive until eight o'clock this morning and by that time the home was just a smoking pile of ash."

"How did Mr. Jackson find out about it so quickly?"

Isaac shrugged. "I have no idea."

"Do you know the name of the family?" asked Nadia.

He shook his head. "I'm not sure. I think he said their last name was either Sodder or Sanders and he gave me the address." Isaac glanced at Nadia. "When we get there you should probably stay in the car until I have time to figure out what's going on."

"No problem, I brought my knitting needles along with me today." She stared out the passenger's side window at the overcast, dreary skies, closed her eyes and offered up a prayer to God for the family who had just lost their precious children. Smoldering timber was still rising from the ashes when Isaac pulled their Chevrolet Clipper alongside the road behind a line of onlookers. He plucked up his notepad and camera, presented his press credentials to the officer in charge of the scene and continued walking toward a crowd of folks who appeared to be investigating the event.

"What happened here?" Isaac asked, after snapping a few photographs.

"This is George and Jennie Sodder's house and apparently some of their children died in the fire last night."

"Have they found the bodies?"

"No," Chief Morris replied. "We've briefly searched the grounds and so far we haven't turned up any trace of human remains. The blaze may have been hot enough to completely cremate the bodies."

"Really?" Isaac asked, with a note of disbelief in his voice. "That's odd. Do they know what started the fire?"

"No, we're assuming faulty wiring."

"What time did it start?" Isaac pulled out his notepad.

"Around one o'clock in the morning," the chief replied.

"What time did the fire department arrive?" Isaac asked to confirm what Mr. Jackson had told him earlier.

"Around eight o'clock," Chief Morris replied dryly.

"It took them seven hours to respond?" Isaac probed.

"Yes, it did."

Isaac murmured, "That's rather strange."

"Meaning what?" the man sarcastically countered.

"I think you know what I mean, sir." Isaac looked him straight in the eyes. "Were Mr. and Mrs. Sodder home when the fire started?"

Chief Morris motioned for him to follow him. They walked several yards away so they wouldn't be overheard. "Look, the family is very upset and confused. All I know is what I've been told."

"I won't take up much of your time, chief. Please just tell me what you know at this point."

"Well, I don't know much. According to the parents, the children had opened a couple presents and around midnight they went to sleep."

"How many children do they have?" Isaac interrupted.

"They have ten children and the next to the oldest boy is away in the Army."

"How many are believed to be dead?"

"Five. Their ages range from five years old to fourteen."

Isaac scribbled this down in his notes. "What happened next?"

"The telephone rang a few minutes later and Jennie, she's the children's mother, ran to answer it before it woke everyone up. Apparently, it was a wrong number. When she was going back to bed, she noticed that all the downstairs lights were still on, the curtains were open and the front door was ajar. She shut off the lights, pulled the curtains and shut the door. She also noticed that one of her older children, Marion, was asleep on the sofa and assumed the other kids were in their beds upstairs. Just as she was falling back to sleep she heard a loud noise, like something hitting the roof and rolling off. A little later she woke up again to find she was surrounded by thick smoke."

"I guess she started looking for the children and ran outside?" Isaac presumed.

"I think she woke George up first then discovered the room he used for his office was on fire and tried to phone the fire department but the phone wasn't working."

"I thought you said someone had telephoned earlier?"

"Yes, but it didn't work when Jennie tried to make a call. So anyway, George frantically scanned the house and ran outside. He looked around the yard and took count of what he knew. He could see his wife holding their youngest daughter, Sylvia, whose crib was in their bedroom, seventeen year old, Marion, and two of his sons, John and George Junior. Marion had been asleep on the couch downstairs and John and George's hair was singed as they escaped the fire from their upstairs bedroom. George instructed Marion to go next door and call the fire department. He figured Maurice, Martha, Louis, Little Jennie and Betty still had to be hiding in the burning house. Evidently, he couldn't see much through the fire and smoke, which had swept through the rooms downstairs and engulfed the staircase in flames."

"So, he couldn't save them," Isaac murmured.

"He kept trying—screaming their names. He tried to reach them through the upstairs windows, by climbing the wall and breaking the window with his arm to get back inside the house. He figured he could prop the ladder, he always kept against the house, up to the top floor to reach the kids but the ladder was missing. Then he rushed over to one of the coal trucks he uses for his business, hoping he could climb on top of it and reach the windows, but it wouldn't start. He tried the other truck but it didn't start either." He shrugged. "George never found them, and…" he pointed toward the house, "that's all that remains."

"How long have they been searching for the bodies?"

The chief pulled his pocket watch from his trousers and

stole a glance. "A few hours now."

"Seriously? And they haven't found anything yet?"

"There's a great deal of rubble to go through, and as you can see things are still smoldering so we have to be careful." He lowered his voice and leaned in toward Isaac. "If the kids were in there, we would have found their bodies by now. I've already told George that there's nothing left of his children's remains." The chief pointed toward a man weeping on the other side of the lawn. "That's George. The poor man's arm was still bleeding when I got here from trying to break the window and his voice is barely audible because he was screaming his children's names, over and over again, as he attempted to rescue them." Chief Morris shook his head remorsefully.

Isaac slowly closed his notepad. "Thank you for the information Chief Morris."

"We're still piecing things together here so something might show up that will help us later," the man responded. "Please be mindful about what you print in your newspaper."

"Yes, sir." Isaac turned to walk away and then redirected his attention back to the chief. "How long did it take for the house to burn down?"

"It took about forty-five minutes for it to burn and collapse."

Isaac's brow furrowed. "So, the family stood here watching their house and children go up in smoke and the fire department didn't arrive until this morning? Is that correct? I just want to make sure I've got the facts correct."

Chief Morris looked chagrinned, ignored his question and turned to walk away.

Isaac stood for a long stretch, watching the fire fighters, investigators and crowd of citizens hovering around the site. He was observing their body language and picking up pieces of whispered conversations going on around him:

"They were immigrants from Italy."

"He's never told anyone why he left Sardinia."

"He shouldn't have been so outspoken
concerning Mussolini."

"They tried to warn him."

Confused, Isaac ambled back to where the car was parked and slid in the driver's seat. He twisted around and stared at Nadia. "This is the oddest thing I've ever heard. It just doesn't add up."

Nadia placed her knitting needles on her lap. "What do ya mean, Isaac?"

He scratched his head thoughtfully. "First off, Chief Morris was acting strange. He couldn't explain why it took the fire department seven hours to respond to the fire when the station is less than two miles away. We passed it driving here. Secondly, he's already determined the bodies of the children will never be found, and this type of fire isn't hot enough to completely cremate a body. They would have to find bones." He pulled his notepad out and flicked through the pages of scribbles. "Then I overheard some people talking in the crowd, one of them said they were immigrants."

"Almost everyone I know is an immigrant," said Nadia.

He nodded in agreement. "I think I heard someone else say that George, that's the father, had been warned and that he shouldn't be so outspoken concerning Mussolini."

Nadia's eyes opened wide. "Are ya kiddin' me? This is America! Do people really think that way?"

Isaac's head dropped down on the steering wheel. "There's another thing, too. The chief indicated the wife, Jennie, had received a call shortly before the fire started. But later, when they tried to call the fire department, the telephone line was dead."

"Who was on the line when she answered the phone before the fire started?"

"Apparently, it was a wrong number. Someone had called

the Sodder's number by mistake."

"Is that the family's name?"

"Yes, George and Jennie Sodder and they have ten children and now five of them are gone."

"Oh, Lordy," Nadia whispered. "This is so tragic."

"Yes, it is. None of it makes sense." Isaac rolled down the car window, snapped a few more shots with his camera and turned to face his wife. "I'm sorry I let you come with me, Nadia. Let me take you home and we'll try to salvage what's left of Christmas Day."

Charleston, West Virginia

1982

I cupped my hand around the steaming coffee mug as the questions dominoed in my mind. "Was the fire started by faulty wiring?" I asked.

Nadia digested this for a moment. "According to Isaac, it was highly unlikely because if the power had been dead then the Christmas lights that were on the tree wouldn't have remained lit as the fire was ravaging their house."

I picked up another donut and took a nibble. "What about the telephone call and then the phone not working when they tried to call the fire department?"

Nadia pointed her finger at me. "That was another interesting part of the story. When the phone rang shortly before the fire started, Jennie answered it and an unfamiliar female voice asked for an unfamiliar name. Jennie heard boisterous laughter and glasses clanking in the background. Mrs. Sodder informed the woman that she had the wrong number and hung up. Soon after the fire started, they tried to call but the phone wouldn't work, Marion ran to the neighbor's house and the neighbor drove into town and called the chief."

"So," I assumed, "the fire had destroyed the telephone lines."

"No." Nadia shook her head. "Later, a telephone repair man told Mr. Sodder that their lines appeared to have been cut, not burned. A person would have had to climb a fourteen-foot pole to cut it."

My jaw dropped open. "Seriously?"

"Yep. A man later admitted he had cut the phone line, thinking it was the power line, but he denied having anything to do with the fire."

"And they believed him?"

"Apparently so." Nadia sighed.

"Why did it take the fire department so long to arrive?"

Nadia leaned in and confided, "Evidentially, Chief Morris said he couldn't drive the fire truck and had to wait for someone else to drive it."

"The *chief* didn't know how to drive the truck?" I echoed. "Did he notify any other firemen or did they just not show up?"

Nadia shrugged her shoulders. "I'm not sure. It was all very suspicious."

"Ya think? It's crazy weird."

I drummed my fingers on the table, deep in thought. "Okay, when George was trying to get into the house while it was burning, things kept going awry—like the missing ladder and his trucks not starting. Did they ever figure out that part of the story?"

"The ladder was later discovered on an embankment more than seventy-five feet away from its usual resting place. George said it had been propped up against the house for months and he never moved it, but that night it was simply gone. Then Isaac later learned that a witness had come forward claiming he saw a man at the fire scene taking a block and tackle used for removing car engines. If this is true, then that is probably why *both* trucks wouldn't start even though he had used them the day before and they worked just fine. The same man, who had admitted to cutting the telephone line, also confessed to the theft of the block and tackle from the property. Still, he insisted he hadn't started the fire." Nadia stood and emptied the remaining coffee into our mugs. "Would ya like me to brew another pot, Dee?"

"No, thanks. I probably shouldn't be drinking this much caffeine in the morning."

She nodded in understanding.

I brought up another point. "Isaac said that Chief Morris had already determined the bodies would never be found, but Isaac told you that this type of fire wasn't hot enough to completely cremate a body, so the bones would surely be found."

"Well," Nadia assured, "they never found the bodies, and when Isaac was taking pictures of the scene he noticed various household appliances in the burned-out basement—still identifiable. He also knew, from covering fires in the past, that bones remain after bodies are burned for two hours at two thousand degrees." She arched her brow. "Their house was destroyed in forty-five minutes."

"I don't understand how five children could perish in a fire and the experts never found *any* remains," I expressed.

"That's the same point that distressed Isaac," Nadia informed me. "So he figured the best way to get the answers he needed was to visit George and Jennie Sodder and talk to 'em in person."

"Did he go back and talk with them?"

"He sure enough did," Nadia assured. "And I went with him."

Peculiar Events

E ven though the bodies were never recovered, on January 2nd a funeral was held for the five Sodder children. The following day, Isaac and Nadia hopped into their Chevrolet Clipper and made the drive to Fayetteville. They knocked on a door of an old farmhouse, where they had been informed they could find George and Jennie, and were ushered inside from the cold.

"Thank you for meeting with us, Mr. and Mrs. Sodder. My name is Isaac and I work for the *Charleston Gazette*, and this is my wife, Nadia."

"Pleased to meet you." Mr. Sodder extended his hand. "You indicated the reason for your visit was that you didn't understand the odd circumstances surrounding the fire, is this correct?"

"Yes, and honestly I am here because when I was at the site of the fire, talking to the chief, and listening to murmurs in the crowd, it occurred to me that there might be foul play involved," Isaac diplomatically explained. "I also realized that you held a funeral service for the children yesterday, although no bodies have been found."

Mr. Sodder hung his head low. "It's true. The bodies have not been located."

"We drove down here today to see if we might be able to

help you discover what really happened."

"Thank you." Mr. Sodder flicked his wrist and guided us into a small living room where Mrs. Sodder was seated on a petite well-worn sofa. It wasn't a fancy room, but it was a comfortable area with wood floors and a tall built-in bookshelf. Nadia joined Mrs. Sodder on the sofa while Isaac and George lowered themselves into two side chairs. "One of our neighbors offered us this house to stay in until we can figure out what to do next," George explained. His attention turned to his wife. "Jennie, this is Isaac and his wife, Nadia. They are here to help us sort through this mess."

Nadia noticed the other woman's eyes were swollen and brimming with tears, so she reached over, gave the woman a gentle squeeze of her hand, and smiled politely.

"Do you mind if I take notes while we talk? It helps me to piece things together when I'm thinking about them later," Isaac asked.

"Not at all."

"Okay." Isaac pulled out his notepad. "Mr. Sodder…"

The man interrupted him. "Please, call me George."

"Yes, sir." Isaac's eyes rose to meet his. "If there is anything you don't feel comfortable talking about or anything that is too upsetting just let me know. I won't be offended."

Jennie Sodder spoke, "We will tell you everything we know. We just want to find out where are children are. We don't believe they are dead! They can't be!"

Isaac bit his lip, not sure of how to start. Finally, he suggested, "Let's start with the fire. It is supposed it was started due to a faulty wiring."

Jennie shook her head fiercely. "No. The Christmas tree lights were still on as the fire consumed the house." Jennie started weeping. "And I heard something hit the roof right before the fire started," she finally muttered.

Nadia gulped.

George explained, "We've been thinking over some past incidents and I remember a visitor came to our house a few months ago looking for work. He told me that the fuse boxes in the back of the house would cause a fire someday. I had just had the house rewired because we installed an electric stove. I will admit his comment concerned me so I called the electric company and they ensured me it was safe."

"Tell them about the life-insurance salesman who visited in October," Jennie sniffled.

A long and mournful groan escaped his throat, setting the stage before he told the story. "A salesman came by the house and I told him I wasn't interested in buying life-insurance. He told me that our house would go up in smoke and our children were going to be destroyed."

"He used the word *destroyed*?" Isaac restated.

"Yes, he did."

Isaac's eyes popped. "Have the authorities spoken to him?"

George sighed long and hard. "Yes, and he said he blamed his remarks on comments I had made against Mussolini."

Nadia noticed George and Jennie exchange a quick glance.

Jennie twisted a handkerchief in her hands. "They didn't find the bodies, you know. The *Fayetteville Journal* reported our children were found, or at least part of a body, but nothing was found."

"I realized this," Isaac confirmed.

"And I've been conducting experiments... " Mrs. Sodder covered her mouth with one hand as though she was embarrassed. "I know it sounds strange but I have been trying to see if I can burn bones, like chicken bones and pork chops bones, to see if they could be devastated by fire and they can't. I'm always left with a heap of charred bones."

George cleared his throat. "We feel that our children are alive. They were kidnapped and the fire was set to cover the crime."

"Kidnapped by whom?" asked Isaac.

"The Sicilian Mafia," George responded frankly.

"Why?"

"I am outspoken concerning Mussolini and the Fascist government of Italy," he paused as a lump formed in his throat. "We are Americans. We have the right to free speech," he added, as though he needed to justify his comment.

"I agree with you, sir. Freedom of speech is part of what makes America great." He eyed the other man. "Would it be possible to visit the site later today?"

"Sure, but even though we were told by the fire marshal to leave the area undisturbed, we couldn't bear looking at it. I bulldozed about five feet of dirt over it and we plan to plant a garden as a memorial."

This time, Isaac and Nadia exchanged a silent but meaningful glance.

Charleston, West Virginia

1982

The first clap of thunder made me jump out of my seat. An approaching storm split the skies open wide and earsplitting lightning and pounding rain came gushing down. I looked out Nadia's kitchen window and saw the skies had darkened. "Wow," I exclaimed, "that startled me."

"The weather man said there would be thunderstorms off and on all day," she informed me. "We need the rain, though."

"Yes, we do," I agreed and settled back into my chair. "Did Isaac continue to keep in touch with the Sodder family? I noticed some of his notes while I was sorting through the articles in the box you lent me."

"Yes, he did. He kept in touch with them for years. George told Isaac that Sylvia, that's their daughter, found a hard rubber object in the yard a few months after the incident. It was a napalm pineapple bomb. You know, the type that is used in wartime. They figured that's what Jennie heard hit the roof that night and what most likely was used to start the fire, and soon after we visited them the sightings started," Nadia explained.

"The sightings?"

"One woman claimed she had seen the missing children peering from a passing car while the fire was still blazing. Another woman, who operated a tourist shop between Fayetteville and Charleston, reported she saw the children the morning after the fire. She said she had served them breakfast and that there was a car with Florida license plates parked in the lot at the same time they were in the restaurant."

"So, the children were alive," I presumed.

"Most likely," Nadia murmured. "Soon after the first two sightings, a woman at a hotel in Charleston telephoned the police and informed them that the children had stayed the night at the hotel. She said two women and two men, who were of Italian origin, accompanied the children. They had rented a large room with several beds. She indicated that when she tried to start a friendly chat with the children one of the men appeared very hostile. He started talking briskly in Italian, then the entire party hushed up and quit talking completely." Nadia shrugged. "The group checked out the next morning and the name they used when checking into the hotel didn't provide any leads either."

I stared out the window watching the rain gush down and felt sorry for the Sodder family even though I didn't know them personally. "Nadia, do you know if they kept searching for the children?"

She snorted. "As far as I know they are *still* searching for their children. Well, Mrs. Sodder is still searching. Mr. Sodder passed away in 1968. If I recall correctly, George and Jennie mailed a letter about the case to the Federal Bureau of Investigation and later received a reply from J. Edgar Hoover stating that the matter appeared to be of a local nature and didn't come within the investigative jurisdiction of the FBI. He did say though, that they would assist if they could get permission from local authorities."

"Did they get permission?"

"Dee, it was unbelievable, but the Fayetteville police and fire departments declined the offer."

I stared at Nadia, dumbstruck. "Seriously?"

She dipped her chin. "Yep. Then George and Jennie hired a private investigator named C.C. Tinsley. He discovered that the insurance salesman who had threatened Mr. Sodder was a member of the coroner's jury that deemed the fire accidental. The investigator also found out that even though Chief Morris had claimed no remains were found, he supposedly confided that he had discovered a heart in the ashes. He said he had buried it inside a box at the scene."

"Why?"

"Who knows? Chief Morris said he hoped that finding any type of remains would pacify the family enough to stop the investigation. However, the private investigator persuaded the chief to show him the spot where it was buried and together they dug up the box and took it to the local funeral director. The funeral director examined the *heart* and determined it was a beef liver that was untouched by the fire."

I regarded Nadia with wide eyes. "That is definitely odd."

Nadia released a long sigh. "Yes, it was. The leads and tips continued to come in for years and George followed up on each one. One time he saw a newspaper photo of schoolchildren in New York and was convinced that one of them was his daughter, Betty. So, he drove to Manhattan but the parents refused to talk to him."

"They wouldn't even talk to him?"

She shook her head. "They wouldn't even let him in the door. Oh, and then some twenty years after the children disappeared, Jennie received a letter in the mail that was postmarked from Kentucky. Inside the envelope was a photograph of a man in his mid-twenties and on its flip side a cryptic handwritten note that read, 'Louis Sodder. I love brother Frankie. IliI. A90132 or A90135.' She couldn't make out the last number and didn't understand what the puzzling code meant, but she did notice that he resembled their son, Louis. Even though he was nine at the time of the disappearance, the chocolate brown eyes, russet wavy hair and same upward tilt of the left eyebrow were enough to convince her that the young man in the photo was her son."

"Were they able to decipher the code?"

"No. Most likely it was a hoax, however the Sodder's were of Italian descent and the number 90132 was, at the time, the postal code for Palermo, Sicily. However, none of the Sodder children were named Frankie, so 'I love brother Frankie' made no sense."

"Did they go to Kentucky to look for him?" I asked.

"They hired another private investigator but never heard back from him."

It was about this time that we heard Jack pounding on the front door. Although the rain had momentarily ceased, he was drenching wet. Nadia opened the front door, only to see him holding a small vintage side-table with spindled legs.

"Where do you want me to put this?" Jack asked, as he shook the water from his loose fitting sandals.

"Out here on the porch, I reckon. What's it for?" Nadia inquired.

"I brought it over for you to fix-up. I also found this bag of doodads." He shoved the sack into her hand.

"Doodads?" Nadia questioned. She peeped into the plastic bag to discover mismatched pieces of jewelry and circular sparkly crystals. She oohed and aahed over small decorative fragments. Nadia was enthralled and her expression spoke for itself. "Are these to spruce up the table?"

Jack looked pleased with himself. "Yeah, I thought you could glue some of them on it like you did with the chair in your living room." He shrugged. "If you think it's a good idea."

"I think it's a great idea, Jack." Nadia smiled brightly. "I can't wait to sort through everything in here." She jiggled the bag. "This is so exciting!"

"I'll help you," I chimed in.

Jack placed the table in the location Nadia had indicated before he entered the house, kicked off his wet sandals, and plopped down on a comfy chair in the living room.

"Nadia was just telling me about the disappearance of the Sodder children," I said to Jack.

Jack ruminated on this for a second. "I remember that tragedy. Did they ever find out what happened to the kids?"

"Not that I know of," Nadia replied. "But Isaac covered the

story for years."

"Really?" Jack's curiosity piqued. "Continue on, Nadia. It sounds like a fascinating tale."

Bones

Charleston, West Virginia
September 26, 1949

"I'm going to a hearing this morning at the capitol," Isaac told Nadia, adding a quick peck on her cheek. "I'm not sure what time I'll be home."

"Is it the hearing about the Sodder case?" asked Nadia.

"Yes, and I know they have been waiting for this for months."

Nadia nodded. She knew it was true. George and Jennie had spoken with Isaac many times over the last few years and they were still desperately searching for answers about their five missing children. "Please tell them I am keeping 'em in my prayers."

"I will," Isaac responded, before striding out the door. He made the drive in less than three minutes and parked alongside the grand West Virginia State Capitol that overlooked the Kanawha River. Even though he had visited the complex many times, Isaac was always awestruck when he glanced at the two hundred ninety-three foot gold dome. He knew it had been gilded in twenty-three karat gold leaf and it always glittered luminously in the morning sunlight. He climbed the stairs, and marveled at the massive amount of white marble when he stepped inside the building.

"Hello," he introduced himself to the guard stationed at the front door. "I am here to attend the Sodder hearing," he explained.

The guard scanned the directory of activities listed on his clipboard and directed Isaac to the tenth door on the left. Upon entering the room, Isaac noticed a sign that read: NO PHOTO-GRAPHS ALLOWED. He had figured as much and hadn't brought his camera with him on this particular morning. He also realized he wasn't too keen on having the Sodders' picture plastered across the front page of newspapers throughout the country— again. Unless, of course, it would somehow help solve the case.

The stately room was set up with straight rows of stiff wooden chairs, an extended oak table lined up against the back portion of the room, and a single chair was placed beside the elongated table. The Sodders sat at a smaller table to the left of the room and Isaac immediately went up to shake Mr. Sodder's hand and added a polite nod to Mrs. Sodder before finding a seat in the second row. Nearly an hour passed before the proceedings began. Isaac slouched in the wood chair with his legs in front of him, crossed at the ankles, and reviewed his previous notes.

Finally, the man overseeing the hearing entered the chamber and took a seat behind the long table. Another man, whom Isaac did not know, seated himself in, what appeared to be designated as, the expert's chair. Isaac straightened, pulled his notebook and pen from his attaché case and listened carefully.

"Please state your name and your involvement in this case," Mr. W.E. Burchett, Superintendent of the West Virginia State Police, requested.

"My name is Oscar B. Hunter and I am a pathologist from Washington, D.C. The Sodder family requested my services to execute an excavation of their property."

Isaac started jotting down notes fast and furiously.

"Did you find anything during the excavation, sir?"

"Yes, there were several items found. We uncovered some damaged coins, artifacts that belonged to the children, a partly burned dictionary and several shards of vertebrae. I then sent the bones to the Smithsonian Institution and they issued the following report." He attempted to hand the report to Mr.

Burchett.

"Please read the result aloud."

"Yes, sir." The pathologist took a long drink of water before he continued. "This report was issued by Marshall T. Newman, who is a specialist at the Smithsonian." Hunter began reading the findings:

"The human bones consist of four lumbar vertebrae belonging to one individual. Since the transverse recesses are fused, the age of this individual at death should have been sixteen or seventeen years. The top limit of age should be about twenty-two since the centra, which normally fuse at twenty-three are still unfused. On this basis, the bones show greater skeletal maturation that one would expect for a fourteen year old boy." He paused for a moment and looked over to Mr. Sodder. "The oldest missing Sodder child was fourteen." He directed his eyes back to the report. "It is however possible, although not probable, for a boy of fourteen years old to show sixteen to seventeen maturation."

Superintendent Burchett asked, "Had the bones been exposed to fire damage?"

"The vertebrae showed no evidence of exposure to fire and it is very strange that no other bones were located during the initial two hour search of the property."

Isaac glanced at Mr. and Mrs. Sodder noticing their faces remained grave and solemn.

The superintendent prompted, "Do you have a theory on why the bones were found if they are not those of one of the Sodder children?"

"Yes, sir. To begin, one would expect to find the full skeletons of five children, rather than only four vertebrae, and since there was no fire damage, it could be assumed that bones were most likely in the supply of dirt Mr. Sodder used to fill in the basement after the fire."

"How could this happen?"

Oscar Hunter didn't skip a beat. "Mount Hope cemetery is located close to the scene of the fire and to the area where Mr.

Sodder collected the dirt to fill in the remains of his home.”

“Is there anything else you would like to add to the report?”

“Yes, sir. I do not believe the vertebrae belonged to any of the Sodder children and I do not believe they perished in the fire.”

“Thank you. That will be all for today.”

Everyone rose to leave, except for Isaac and Mr. and Mrs. Sodder. He walked over to meet them. “I’m sorry this didn’t provide any new information but I’m glad it has been determined your children did not perish in the fire.”

Mrs. Sodder’s eyes looked at him bewildered. “Yes… yes, but where does that leave us?” She closed her eyes tightly. “How are we going to find them?”

Isaac’s heart ached for them. “Please let me know if I can be of any assistance. I’ll do anything you ask.”

“Thank you,” George Sodder shook his hand. “We appreciate you coming today.”

Isaac offered a half smile before leaving George and Jennie alone to process the newly presented information about the dreadful case.

Charleston, West Virginia

1982

"What became of the hearings? Was anyone able to help George and Jennie?" I asked.

"No, in 1950 the West Virginia Legislature closed the case at the state level and declared it hopeless."

I sighed. "That's so sad."

Jack agreed. "All I remember is hearing a few things from time to time over the years. I always hoped they would find out what happened to the children, if for no other reason than the parents' peace of mind."

"Me too." Nadia tilted her head to one side. "Anyway, that's all I remember about the Sodder case. Well, other than George and Jennie erected a billboard with photographs of the children along Route 16 and passed out flyers, too. If I remember correctly, they offered a reward of five thousand dollars and then upped it to ten thousand dollars. After this, they did receive a letter from St. Louis saying Martha—she was the oldest girl—was living in a convent there, and someone in Florida suggested the children were living with a distant relative of Jennie's. I think a man from Texas even contacted them stating he had overheard an incriminating conversation in a bar about the fire. George investigated every lead but always returned home without answers."

"I think there is a photograph of the billboard in one of Isaac's articles," I told Nadia. "I believe I saw it last night." I sprinted to the kitchen and grabbed the box. Easily locating the article, I waved the newspaper clipping in the air. I leaned in closer, squinting my eyes. The huge white billboard featured portraits of each of the missing children. They all had dark hair, and even through the shades of gray, I could see the sparkle in their eyes. *Why are there six portraits on the board? Only five children disappeared.* The additional one, I realized was a replica of the photo Jennie received in the mail—the one she believed was Louis, after he was grown. "Look! It says, AFTER THIRTY YEARS IT IS NOT TOO LATE TO INVESTIGATE."

I handed it to Nadia and she stared at the photograph of the billboard for a long moment before she tendered it to Jack. I noticed her hand lingered on his a little longer than necessary.

"This picture sure does tell a story, don't it?" Jack commented.

I agreed wholeheartedly, "It sure enough does."

"I can't read the words on the bottom portion. Can you see it well enough to read it to me, Dee?" Jack offered it to me with an outstretched hand.

I snatched it up and read out loud:

ON CHRISTMAS EVE 1945 OUR HOME WAS SET AFIRE AND FIVE OF OUR CHILDREN AGES FIVE THOUGH FOURTEEN KIDNAPPED. THE OFFICIALS BLAMED DEFECTIVE

WIRING ALTHOUGH LIGHTS WERE STILL BURNING AFTER THE FIRE STARTED.

THE OFFICIAL REPORT STATED THAT THE CHILDREN DIED IN THE FIRE HOWEVER NO BONES WERE FOUND IN THE RESIDUE AND THERE WAS NO SMELL OF BURING FLESH DURING OR AFTER THE FIRE.

WHAT WAS THE MOTIVE OF THE LAW OFFICERS IN-VOLVED? WHAT DID THEY HAVE TO GAIN BY MAKING US SUFFER ALL THESE YEARS OF INJUSTICE? WHY DID THEY LIE AND FORCE US TO ACCEPT THOSE LIES?

*PICTURE NO 6 RECEIVED IN 68 LOUIS ONE AND THE SAME/NOW IN ANOTHER STATE.

I placed the article back into the box. "So Nadia, what was Isaac's theory on the case?"

"He figured that the Sicilian Mafia tried to recruit George and he had refused to join 'em."

"What about the fire and the kidnapping?"

"Well, it is possible the children were kidnapped by some-one they knew. Perhaps, the person entered the front door, told the kids about the fire and promised they would take 'em some-where safe."

"Then why didn't the children attempt to contact their par-ents?" I asked.

Nadia suggested, "They may have been murdered after they were kidnapped or perhaps they were told that their parents and siblings would be harmed if they tried to get in touch with them. Who knows? Stranger things have happened."

I gave her a doubtful look. "I'm not sure if stranger things have happened... because that's probably the most bizarre story I've *ever* heard."

Jack rubbed his chin thoughtfully. "It is mighty odd. Folks have been tryin' to solve this case for decades now, and though the disappearance of the Sodder children will likely remain a mystery forever, I doubt the incident will ever be forgotten."

"Perhaps they were kidnapped and sold on the black market like the children who were abducted by Georgia Tann," I suggested.

"Who is Georgia Tann?" asked Jack.

"She was the director of a Memphis based adoption organization from the mid-1920's until 1950. She kidnapped children and sold them to her wealthy patrons. Several famous people adopted children through The Tennessee Children's Home Society. Celebrities like Joan Crawford, June Allyson and her husband, Dick Powell, and one of the governors of New York adopted a child from Tann, too. But they weren't aware of the devices she used to obtain many of the children. I read somewhere that hundreds of children died at the institution due to abuse and neglect."

"Really?" Jack scrunched his nose in disgust.

"Possibly..." Nadia pondered on this for a long moment. "I only know *fragments* of *what* happened, I don't have a clue as to *why* it happened."

We sat in companionable silence, each of us most likely deliberating on the fate of the five Sodder children until Jack abruptly blurted out, "We may as well spill the beans, don't you think, Nadia?"

She nodded in agreement.

"Spill the beans about what?" I questioned.

Jack cleared his throat. "Nadia accepted my proposal."

"That is wonderful!" I exclaimed. I was delighted Nadia was going to work at the pawn shop, or at least refurbish some furniture. I knew she could use the extra money and it would give her something to fill her time. "Are you going to pay her by the job or by the hour?"

"Pay her?" Jack uttered, when I lifted my eyebrow at him.

"Surely you don't expect her to do it without getting paid," I huffed.

Jack startled and I noticed Nadia's eyes were soft and wistful when she glanced at him. Bing twittered gently in the background, as though setting the stage for the upcoming declaration. I stared at Jack anticipating his response.

He clarified, articulating each word deliberately, "I asked

Nadia to marry me, Dee."

"Oh…" I shifted in my seat, a grin tugging at my cheeks.

"Stop smirking," Jack and Nadia said at the same time.

"Who, me?"

"Yes, you!"

I chuckled under my breath. *Like I didn't see this coming.*

Epilogue

N
adia and Jack tied the knot on a crisp, cold day in December 1982. They had a small ceremony, which was held at the First Baptist Congregational Church, and since I was in charge of coordinating the event I am happy to report that everything went off without a hitch.

They ultimately turned the pawn shop into an upscale resale store, with beautiful displays in the window highlighting Nadia's painted furniture. I was also present on the day when they hung the new motif over the shop on Capitol Street. Dozens of people gathered around.

The elaborate diorama was a festive celebration of Valentine's Day from mirrors surrounded by a decorative mosaic made from dishes, marbles, shells and bits of broken tiles, to swirling antique games and signs with uplifting words like LOVE and AMAZING and YOU ARE AWESOME! These were all staunchly positioned on chairs and tables painted in bright vibrant shades of yellow, pink, and blue. Boxing memorabilia was tastefully featured in the corner with a simple phrase that beckoned, "DO YOU NEED A GIFT FOR YOUR MAN?"

Jack and Nadia stood side-by-side, holding hands, while the *Jack and Nadia's Fine Antiques and Collectibles* sign was hoisted high and placed above the bright white French door of the entryway. I savored the moment, so I waved them over to the perfect spot.

"Say cheese!" I took a snapshot. *How could I not?*

After all, "Every picture tells a story—don't it?"

"Sometimes you will never know the true value of a moment until it becomes a memory."

— *Dr. Seuss*

Author's Notes and Reading List

Although the Benwood Mine Disaster took place on April 28, 1924, the last body was not recovered until May 6th. It took eight days for the rescuers to find the bodies of all the men and boys who were killed in the mine. The actual number of men and boys who died may never be known, but it was reported that the explosion claimed the lives of 119 coal miners. There were no survivors. It is the third worst coal mining disaster in the state of West Virginia.

https://en.wikipedia.org/wiki/Benwood_mine_disaster
https://usminedisasters.miningquiz.com/saxsewell/benwood.htm

According to Archiving Wheeling, miners who were burned beyond recognition were buried in a mass grave at Greenwood, which is some seventy-two miles south. This article also confirms that a grieving widow tried to drown herself in the Ohio River after learning of her husband's death. Another woman, in the crowd of onlookers, was struck and killed by a speeding truck rushing supplies to the rescue teams. This article also verifies that a relief fund was established for widows and children of the miners at the Bank of Benwood. Two employees embezzled the money, which was never recovered. The embezzlers were caught and convicted, receiving ten-year sentences in the West Virginia State Penitentiary at Moundsville. Damaged by the scandal, the Bank of Benwood closed in 1925.

http://www.archivingwheeling.org/blog/no-survivors-91-years-ago-today-in-benwood

Many of the details concerning the rescue efforts are documented in the West Virginia Archives and History. This article states there were 111 men and boys lost in the mine and includes an official list of those who were known to be in the mine at the time of the accident.
http://www.wvculture.org/history/disasters/
benwood03.html

According to the West Virginia Encyclopedia, two fire bosses had reported the mine free of gas before the workers entered the portal to begin their shift. A miner found a roof fall about 22 feet from the room face. Thinking the fall had been examined, he went over the fall toward his room and his open light ignited the afterdamp, an explosive mixture of methane and air. In all probability the fall occurred after the fire boss had visited the area. The mine was dry and dusty—sprinkling and ventilation were poor—and the subsequent dust explosion carried to every area of the mine.
https://www.wvencyclopedia.org/articles/450

The Benwood Mine Disaster Memorial, which is three large stones, commemorates the coal miners killed in the April 28, 1924, explosion. The monuments that now stand at the Boggs Run Road site are the result of the generosity of numerous individuals and organizations.
http://www.benwoodwv.com/minedisaster.htm

In 1924, the same year of the Benwood Mine Disaster, Congress proposed a constitutional amendment prohibiting child labor, but the states did not ratify it. Then, in 1938, Congress passed the Fair Labor Standards Act. It fixed minimum ages of sixteen for work during school hours, fourteen for certain jobs after school, and eighteen for dangerous work.
https://www.scholastic.com/teachers/articles/teaching-content/history-child-labor/

I was able to validate the story told by Mrs. Ruth Carrington concerning the lynching of Anthony Crawford. The Internet link can be found here:
https://abhmuseum.org/anthony-crawford/

The flood Nadia spoke of occurred on March 19, 1936. At the time, they were living in Wheeling, West Virginia. They lost many irreplaceable items in the flood. Fortunately, their family was not harmed. However, most of their beloved photographs were ruined when the floodwater swept through the first floor of their home.
http://www.gendisasters.com/west-virginia/20575/wheeling-wv-flood-mar-1936

To read more about school integration in West Virginia, please visit the following webpages:
http://www.wvculture.org/history/education/trent.html
http://www.wvculture.org/history/africanamericans/schoolintegration004.html

George Sodder died in 1968. Jennie Sodder continued to search for her missing children, although she never found them. For the rest of her life, she wore black in mourning and tended to the memorial garden at the site of the former house that was destroyed by fire. She died in 1989. After her death, the family removed the weathered, worn billboard. What happened to the Sodder children? This remains an unsolved mystery.

George Sodder's gravesite and other photographs can be found at:
https://www.findagrave.com/memorial/18304007/george-sodder

If you would like to learn more about the disappearance of the Sodder children or see their photographs visit:
https://www.smithsonianmag.com/history/the-children-

who-went-up-in-smoke-172429802/

Books

Melody Bragg and George Bragg, *West Virginia Unsolved Murders and Infamous Crimes.* Glen Jean, WV: GEM Publications, 1993.

"Missing or Dead?" *Greensboro News and Record,* November 18, 1984.

Photo Credits

Photograph of Nadia in a straw hat by LuliiaVerstaBO at Deposit Photos.
Photograph of Sodder Children Billboard from www.appalachianhistory.net

I sincerely hope you enjoyed reading *Appalachian Tales*. I would greatly appreciate your feedback with an honest review on Amazon.com or Goodreads. First and foremost, I'm always looking to grow and improve as a writer. It is reassuring to hear what works, as well as to receive constructive feedback on what should be improved. Secondly, proceeds earned from this book are donated to the Monroe County Humane Society, and the animals can always use your help.

Best regards,
Deanna Edens

CPSIA information can be obtained
at www.ICGtesting.com
Printed in the USA
LVHW031645220320
650836LV00019B/3269